Rheinwerk Computing

The Rheinwerk Computing series offers new and established professionals comprehensive guidance to enrich their skillsets and enhance their career prospects. Our publications are written by the leading experts in their fields. Each book is detailed and hands-on to help readers develop essential, practical skills that they can apply to their daily work.

Explore more of the Rheinwerk Computing library!

Michael Kofler

Scripting: Automation with Bash, PowerShell, and Python

2024, 470 pages, paperback and e-book
www.rheinwerk-computing.com/5851

Metin Karatas

Developing AI Applications: An Introduction

2024, 402 pages, paperback and e-book
www.rheinwerk-computing.com/5899

Philip Ackermann

Full Stack Web Development: The Comprehensive Guide

2023, 740 pages, paperback and e-book
www.rheinwerk-computing.com/5704

Bernd Öggl, Michael Kofler

Git: Project Management for Developers and DevOps Teams

2023, 407 pages, paperback and e-book
www.rheinwerk-computing.com/5555

Bernd Öggl, Michael Kofler

Docker: Practical Guide for Developers and DevOps Teams

2023, 491 pages, paperback and e-book
www.rheinwerk-computing.com/5650

www.rheinwerk-computing.com

Michael Kofler, Bernd Öggl, Sebastian Springer

AI-Assisted Coding

The Practical Guide for Software Development

Editor Megan Fuerst
Acquisitions Editor Hareem Shafi
German Edition Editor Christoph Meister
Translation Winema Language Services, Inc.
Copyeditor Julie McNamee
Cover Design Graham Geary
Photo Credits Midjourney.com; Shutterstock: 1740319322/© Iurii Motov
Layout Design Vera Brauner
Production Kelly O'Callaghan
Typesetting SatzPro, Germany
Printed and bound in Canada, on paper from sustainable sources

ISBN 978-1-4932-2693-1
1st edition 2025

© 2025 by:
Rheinwerk Publishing, Inc.
2 Heritage Drive, Suite 305
Quincy, MA 02171
USA
info@rheinwerk-publishing.com
+1.781.228.5070

Represented in the E.U. by:
Rheinwerk Verlag GmbH
Rheinwerkallee 4
53227 Bonn
Germany
service@rheinwerk-verlag.de
+49 (0) 228 42150-0

Library of Congress Cataloging-in-Publication Control Number: 2024057967

Contents at a Glance

Contents

2   Pair Programming

3   Debugging

4   Refactoring

7   Databases   195

8   Scripting and System Administration   221

PART II   Local Language Models and Advanced AI Tools

9   Executing Language Models Locally
253

10   Automated Code Processing

11   Level 3 Tools: OpenHands and Aider   313

12   Retrieval-Augmented Generation and Text-to-SQL   337

13   Risks and Outlook   367

Preface

You've almost certainly already asked ChatGPT or another artificial intelligence (AI) tool a coding question and saw the tool spit out (seemingly) perfect code within seconds.

This book explores the potential and limitations of AI tools in software development. Let's start with the positives. We'll show you that ChatGPT, GitHub Copilot, and others can do more than just generate a few lines of code. You can also use them to create unit tests, search for errors, perform refactoring tasks, develop database schemas, optimize SQL queries, administer servers, and write scripts.

But the brave new AI world also has its downsides. Even if the code looks plausible and the accompanying text is elegantly formulated, some AI suggestions are simply wrong due to logic errors, the use of nonexistent functions, or references to variables whose names are actually different. Although the code does work sometimes, it's inefficient or insecure.

AI tools make mistakes—that much can't be disputed. Used correctly, however, these new tools can also save you a lot of time. That's what this book is all about.

Local Execution of Language Models

Companies in particular often have problems entrusting their code base to tools that run in the cloud. Even if AI providers promise not to use your code for the training of future language models in company offers, the question arises as to how far you can trust these promises and whether data protection problems may arise.

The easiest way out is to run the language models required for AI tools locally. You can't download the large language model (LLM) from ChatGPT, but various alternatives are available free of charge that work similarly well.

However, the local execution of LLMs consumes a lot of resources, so only powerful notebook computers or PCs are suitable for this. New CPUs with neural processing units (NPUs) promise new possibilities in this respect. However, a real benefit for the local execution of LLMs only arises in combination with suitable software—and there is still a lot to be done.

There is one more option: you can set up your own server in your company or in a data center that is available to all employees for AI tasks. In that case, "local" doesn't refer to your own notebook computer, but to a company-owned computer. In this book, we'll show you how you can use local language models and even optimize them with your own training material.

We'll also present the *GPT4All*, *Ollama*, and *Tabby* programs to run free LLMs locally. You can then use the local language models in chat mode for coding (e.g., by using the *Continue* editor plug-in) or via an application programming interface (API) to develop your own AI applications.

Coding with AI Support for Advanced Users

If you've asked ChatGPT for advice on a coding problem or if you've followed suggestions from GitHub Copilot (or a similar tool), then you know about interactive AI support for coding. But there's more to it! You can go one step further by using the appropriate libraries:

- **Processing many code files automatically**
  You want to change all comments in 100 files from German to English? You want to migrate a project from Python 2 to Python 3? Instead of interactively processing file by file, couldn't you write a script that automates this task? We've tried it out and will share our experiences with you.

- **Level 3 tools**
  Similar to self-driving cars, there are also categories of autonomy for AI coding tools. ChatGPT, Claude, GitHub Copilot, and other tools are assigned to levels 1 and 2 depending on their application (*code completion* or *code creation*). However, the first level 3 tools (*supervised automation*) already exist that go beyond this and independently generate or modify code files according to your instructions. We experimented with *OpenHands* and *Aider*, and we were particularly enthusiastic about the Aider concept.

- **Retrieval-augmented generation (RAG)**
  It's currently impossible for private users or small businesses to train their own language models. By using RAG, however, you can add your own data to language models. This results in a host of additional application options, including the integration of daily updated data or the automated conversion of text prompts into SQL queries. In this chapter, we focus primarily on the *LlamaIndex* library.

Theoretically, the tools outlined here represent the next development step in the use of AI in coding. In real life, however, our experiences have been mixed. Not everything worked out as we had hoped. AI is already able to perform some of these tasks, but for more complicated requests, patience is still required until these tools become suitable for everyday use.

AI Makes Mistakes and Hallucinates

If you talk to other developers, you'll hear many anecdotes about the major mistakes AI tools have made recently. There are also many chapters in this book in which we show examples with faulty AI code, incorrect conclusions, or simply invented functions or options. (Invented functions have to do with an unpleasant feature of current language models: If AI can't give a reliable answer due to insufficient training material, it invents a text that sounds as plausible as possible—that is, AI "hallucinates.")

Depending on how you use AI tools, the error rate varies between rather low (widespread programming languages, everyday problems) and quite high (exotic languages, new libraries/APIs, very specific tasks). The biggest problem isn't that mistakes happen. We all know that people also make mistakes, probably even more often. What is annoying is that AI tools argue even the biggest nonsense politely and at a pretty high linguistic level. That's what makes it so difficult to tell the right answers from the wrong ones. In our view, AI tools will only be truly "intelligent" when they answer: "I don't know" or "I'm not sure about this proposed solution, but it could work like this." We aren't that far yet in early 2025, even if there are efforts to show how the answers can be derived.

As you can see, we aren't trying to sell you an AI tool! On the contrary, we point out the issues we've encountered in all chapters, especially in Chapter 13.

The Three Stages of Using AI

While writing this book, we not only worked intensively with AI tools ourselves but also spoke to many other developers. Through this, we've recognized a pattern in the use of AI tools in coding. It corresponds fairly closely to the *hype cycle*, a model coined by consultant Jackie Fenn to describe how new technologies are perceived and used (see Figure 1):

- **Euphoria**
  The first experiments will leave you, like most developers, speechless with enthusiasm. This "thing" understands your questions, gives answers that make sense, and even explains the code. That's crazy!

- **Disillusionment**
  Then, the moment arrives when the AI tool of your choice makes its first mistakes. It's no longer about Hello World functions, but about real, complex code. The AI tool explains the world to you with the utmost confidence, and you realize that the explanations are convincing in terms of language, but completely wrong in terms of content. You might conclude here that AI tools are useless, their use irresponsible, and so on.

- **Pragmatic use**

  After a few weeks, you'll develop a good feeling for when AI results are most likely correct and where you need to make improvements. You're getting better and better at assessing the limits. What remains is a tool that makes mistakes, but is still a real help within the scope of its capabilities. You question every AI answer and reckon with an error rate of perhaps 10%. But you use what works—and that's quite a lot.

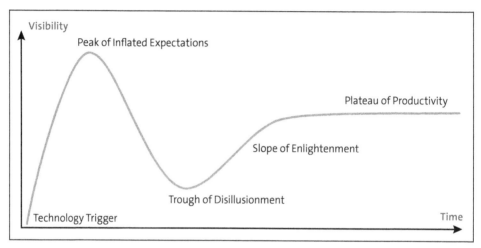

Figure 1 AI Tools Currently in the Stages of the Hype Cycle

There's No Getting around AI Tools

Despite all the problems, we're convinced that AI tools are here to stay. According to one study (*https://www.atlassian.com/software/compass/resources/state-of-developer-2024*), managers believe that the time-saving factor is huge, while developers who actually use AI tools are much more cautious (and realistic) about the benefits!

The extent of the efficiency gain depends heavily on the type of code you develop. But there is always a certain amount of time saved and a higher level of convenience at work. Regardless of whether you see the glass as half full or half empty, coding using AI support increases your efficiency today and will increase it even more tomorrow.

Chemist Derek Lowe put it this way in February 2023:

> *I don't think AI and automation will replace chemists, but rather that the human chemists who use them well will replace the ones who don't.*

We're convinced that this statement, which has often been quoted since, also applies to the IT world. AI therefore won't replace software developers, at least not in the near future. But developers who use AI tools in a reasonable way will replace their colleagues who do not.

Now is the right time to get to grips with AI tools! Yes, AI tools will become even better in the coming years, but the potential is already huge. And the knowledge and experience you gain today will help you better understand new AI technologies tomorrow and integrate them more quickly into your day-to-day work.

According to the 2024 Developer Survey (*https://survey.stackoverflow.co/2024/ai*), 63% of all developers already use AI tools. An additional 13% are planning to use them in the near future. We're clearly at the beginning of a new era in terms of software development. Despite all the fears associated with such far-reaching changes, we've found the use of AI tools in our day-to-day coding work very satisfying. AI tools haven't done the thinking for us, but they have done a lot of boring typing. The concepts of the software we developed still came from us, but the often-tedious work of implementation has become easier for us.

With this in mind, we hope you enjoy reading this book. Be inspired by our examples, and look for ways to integrate AI tools into your day-to-day work in a meaningful way. Take advantage of the opportunities that arise here!

Michael Kofler (*https://kofler.info*)
Bernd Öggl (*https://webman.at*)
Sebastian Springer (*https://sebastian-springer.com*)

PS: This book deals exclusively with the use of AI tools for software development and administration. Clearly, the new developments in the AI sector will have a wide-ranging impact on society, both positive and negative. However, this isn't a topic in this book, nor is the use of AI tools outside of IT.

Sample Files

There are sample files for a few chapters of this book, which you can download here: *www.rheinwerk-computing.com/6058*.

PART I

Coding with AI Support

Chapter 1

Introduction

This Hello World! chapter provides an initial introduction to the use of artificial intelligence (AI) tools in program development. With examples, the chapter shows the use of chat-based tools such as ChatGPT and coding wizards such as GitHub Copilot.

You've probably already tried out AI coding tools. Nevertheless, you should skim the following pages to get some new ideas in the introductory chapter before we go into more depth in the later chapters. For example, Section 1.5 contains a concise summary of how language models work behind the scenes. This knowledge will help you better understand the limits of current technology.

1.1 Coding via Chat

Chat systems such as ChatGPT provide the most intuitive introduction to AI-supported coding. When we wrote this book, there were various free systems to choose from:

- ChatGPT (the original from OpenAI; *https://chatgpt.com*)
- Copilot (Microsoft using ChatGPT/OpenAI; *https://copilot.microsoft.com*)
- Gemini (Google; *https://gemini.google.com*)
- Claude (Anthropic; *https://claude.ai*)
- DeepSeek (the disrupting newcomer from China; *https://chat.deepseek.com*)

On most websites, you have the option to activate additional functions by purchasing a paid subscription. To the extent that we experimented with publicly available chat systems while working on this book, ChatGPT was our preferred system. You must register once on this website, but that's quick and doesn't cost anything. Then, the application is very simple: you formulate your question and receive an answer.

1.1.1 Password Generator

You need a function in a Python script that returns a random password of a specified length. Of course, you can write the code yourself, but you'll probably save yourself a minute or two if you hand over this trivial task to ChatGPT.

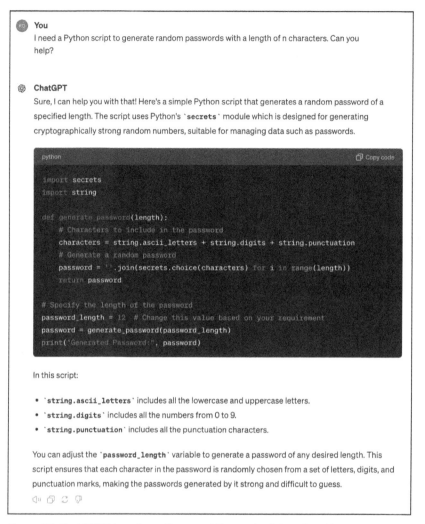

Figure 1.1 ChatGPT Wrapping a Password Generator into a Function

You can now easily copy and paste the code into your script. Don't forget the necessary imports! These are always specified at the beginning of the script.

1.1.2   Python: Searching and Replacing Dates

In this example, all dates in a character string are supposed to be converted from the US format (12/31/2024) to the ISO format (2024-12-31). The programming language to be used is Python.

The code uses the re module, which is used to process regular expressions in Python. ChatGPT formulates the code as a function, which simplifies further use. convert_us_to_iso_date in turn contains a subfunction named replace, which gets passed as a parameter to re.sub and replaces each hit with the newly formatted date.

Prompt: Python: I want to replace all US dates by ISO dates in a string. Can you help?

Certainly! To replace all US date formats (commonly MM/DD/YYYY) with ISO date formats (YYYY-MM-DD) in a string using Python, you can use the `re` module to find dates and then rearrange the format. Here's a Python function that accomplishes this:

```python
import re
def convert_us_to_iso_dates(text):
    # Pattern to find dates in MM/DD/YYYY format
    date_pattern = r'\b(\d{1,2})/(\d{1,2})/(\d{4})\b'

    # Function to replace each match
    def replace(match):
        month, day, year = match.groups()
        return f'{year}-{month.zfill(2)}-{day.zfill(2)}'

    # Replace all matches in the text
    return re.sub(date_pattern, replace, text)

# Example usage
input_text = "Today's date is 05/09/2024 and the event " + \
             "was on 12/25/2023."
converted_text = convert_us_to_iso_dates(input_text)
print(converted_text)
```

This script:

- Defines a regex pattern to match dates in the format MM/DD/YYYY.
- Uses a replacement function to rearrange the date components into the ISO format YYYY-MM-DD.
- Applies this transformation to all occurrences in the provided text.

The code works perfectly in the form presented here. If you're not familiar with regular expressions, you can have the structure of the pattern explained to you in a follow-up question:

Prompt: Could you explain the regular pattern?

Certainly! Here's a breakdown of the pattern `r'\b(\d{1,2})/(\d{1,2})/(\d{4})\b'`:

- `\b`: This is a word boundary marker. It ensures that the pattern matches whole words only, which helps prevent partial matches of larger sequences that aren't dates.

- (\d{1,2}): This part of the pattern captures one or two digits. The \d denotes a digit (equivalent to [0-9]), and {1,2} specifies that the digit can appear once or twice. This pattern is used for both the month and the day in the date.
- /: This is a literal forward slash character. It separates the month and day, and day and year, in US date formats.
- (\d{4}): This captures exactly four digits, representing the year. Here, \d again denotes a digit, and {4} specifies that exactly four digits must be present.
- \b: Another word boundary marker, ensuring the date is a standalone entity, not part of a larger sequence of numbers or characters.

So, the entire pattern matches strings that start and end with a boundary between words (like spaces or punctuation), followed by one or two digits for the month, a slash, one or two digits for the day, another slash, and finally four digits for the year.

In this book, we normally refrain from reproducing such explanations or abbreviate ChatGPT outputs if they are too verbose because we're primarily interested in how you can best formulate the prompt.

ChatGPT usually uses variable names composed with an underscore (convert_us_to_iso_dates) in Python code. If you prefer other conventions or if other rules apply in your project, you can specify these directly in the prompt or in a follow-up question:

Prompt: Could you please regenerate the code with camelCase names?

Sure! Here's the Python code updated to use camelCase naming conventions for the function and variable names:

```python
def convertUsToIsoDates(text):
    # Pattern to find dates in MM/DD/YYYY format
    datePattern = r'\b(\d{1,2})/(\d{1,2})/(\d{4})\b'
```

AI Results Aren't Reproducible!

If you enter the prompts printed previously, you'll receive different answers even if you're using exactly the same ChatGPT version. Current AI tools produce new text or code each time, so the result might be better or worse.

1.1.3   SQL: Analysis of a Table

The starting point for the following example is a table in MySQL that contains value-added tax (VAT) rates (sales tax rates) for different countries. As the tax rates can change over time, the validFrom column indicates the date from which the value is valid. What does the SELECT query look like that returns the tax rate valid at a certain time for a certain country?

For ChatGPT to give the correct answer, you must provide sufficient context information. In particular, it's important which database management system (DBMS) you're using (in this case, MySQL) and what the exact structure of the table is.

> **Prompt:** MySQL: I have a table to store VAT rates for different countries. This is the table definition:
>
> ```
> CREATE TABLE vat (
>     id       INT NOT NULL AUTO_INCREMENT,
>     country  CHAR(2) NOT NULL,
>     rate     DOUBLE NOT NULL,
>     validFrom DATE NOT NULL,
>     PRIMARY KEY (id)
> )
> ```
>
> I want to get the VAT rate for a country (i.e., 'de') at a given time ('2024-04-15'). What should the query look like?

To retrieve the VAT rate for a specific country at a given time from your table, you need to write a SQL query that checks for the most recent rate that has been set before or on the date you're interested in. This involves selecting the correct entry based on the country code and the validFrom date.

Here is the SQL query to find the VAT rate for Germany ('DE') as of April 15, 2024:

```
SELECT rate
FROM vat
WHERE country = 'DE'
  AND validFrom <= '2024-04-15'
ORDER BY validFrom DESC
LIMIT 1;
```

ChatGPT provides the correct answer. The program then explains the structure of the SQL command in detail (not printed here). If you don't use SQL on a daily basis, this is a great help in understanding the code.

Note that you must explicitly state which database system you're using for questions of this type. LIMIT 1 is a MySQL-specific extension of the SQL syntax. The same question for SQL Server provides a different answer with TOP instead of LIMIT.

The only change needed to make it fully compatible with SQL Server is to replace LIMIT 1 with TOP (1), which is the SQL Server syntax for limiting the number of rows returned. Here's the revised query for SQL Server:

```
SELECT TOP (1) rate
FROM vat
WHERE country = 'DE'
  AND validFrom <= '2024-04-15'
ORDER BY validFrom DESC;
```

This shows us how important it is to specify the context of the question.

1.1.4   PHP: Bearer Token Authentication

Say you're familiar with PHP and the concepts of REST application programming interfaces (APIs), but you don't know how to evaluate a bearer token in a PHP program. No problem, ChatGPT can help you out:

In PHP, when handling RESTful API requests that use a bearer token for authentication, you typically extract the token from the Authorization header of the incoming GET request. Here's a step-by-step guide on how to do this:

- Access the authorization header: PHP provides access to request headers through the $_SERVER superglobal array. The Authorization header can be accessed using the key HTTP_AUTHORIZATION within this array.
- Extract the bearer token: The value of the Authorization header generally starts with the word "Bearer", followed by a space and then the token itself. You'll need to parse this string to extract the token.

Here's an example code snippet that demonstrates how to extract a bearer token from a GET request:

```php
<?php
// Function to get the bearer token from the
// Authorization header
function getBearerToken() {
    if (isset($_SERVER['HTTP_AUTHORIZATION'])) {
        // Extract the header value
        $header = $_SERVER['HTTP_AUTHORIZATION'];
```

```
    // Check if the header starts with "Bearer"
    if (preg_match('/Bearer\s(\S+)/', $header, $matches)) {
        // Return the token part
        return $matches[1];
    }
  }
  return null; // Return null if no token was found
}
?>
```

Handle edge cases:

- **Header Not Present**: If the `Authorization` header is not present, your function should handle this gracefully, perhaps returning `null` or an appropriate error message.
- **Malformed Authorization Header**: If the header is present but doesn't properly include a bearer token (e.g., missing the "Bearer" prefix), you should also handle this case, possibly logging the issue or returning an error.

Always ensure that tokens are handled securely:

- Validate the token if necessary (e.g., check its format, expiration, and authenticity).
- Use HTTPS to protect the token in transit.
- Limit the exposure of sensitive token information in logs or error messages.

This approach will help you securely extract and handle a bearer token in a PHP-based RESTful API environment.

1.1.5 Porting a Bash Script for PowerShell

ChatGPT can do more than just generate new code. It can also search for errors in given code; convert code from one programming language to another; develop unit tests for a given function; and restructure, comment, and explain existing code.

The last introductory example is about porting a simple Bash script for PowerShell. The original script monitors the Markdown files in a directory. When changes are saved in a file and the Markdown file is newer than the corresponding PDF file, the PDF document is recreated using the `pandoc` program. This automated build process is now to be implemented on a Windows computer using PowerShell. That's no problem for ChatGPT!

Prompt: Can you provide a PowerShell equivalent to the following Bash script?

```
#!/bin/bash
while true; do
  for mdfile in *.md; do
    pdffile=${mdfile%.md}.pdf
```

```
    if [ $mdfile -nt $pdffile]; then
      echo $mdfile
      pandoc ... $mdfile -o $pdffile
    fi
  done
  sleep 1
done
```

Certainly! Here is the PowerShell equivalent of the given Bash script. This script loops continuously, checking for .md files in the current directory. If an .md file is newer than its corresponding .pdf file, it uses Pandoc to convert the Markdown file to a PDF. It also waits for 1 second after each iteration.

```
while ($true) {
    Get-ChildItem -Filter *.md | ForEach-Object {
        $mdFile = $_
        $pdfFile = "$($mdFile.BaseName).pdf"

        if (-Not (Test-Path $pdfFile) -or `
            (Get-Item $mdFile).LastWriteTime -gt `
            (Get-Item $pdfFile).LastWriteTime)
        {
            Write-Output $mdFile
            pandoc $mdFile -o $pdfFile
        }
    }
    Start-Sleep -Seconds 1
}
```

Make sure to customize any Pandoc-specific parameters according to your conversion requirements.

ChatGPT "understands" the task of the original script, even though it doesn't contain any comments. The new code not only uses PowerShell cmdlets but also takes into account the fact that PowerShell must ensure both files really exist before comparing the modification date of two files. Great!

The answer provided by ChatGPT contains a step-by-step explanation of the new script, which we haven't reproduced here for space reasons. We've also had negative experiences with this type of translation task, regardless of the programming language: from time-to-time ChatGPT invents commands, cmdlets, or functions that don't even exist.

1.1.6 Paid ChatGPT Offers

ChatGPT is basically free. If you're willing to spend $20 per month, ChatGPT Plus gives you access to the latest language model, more resources, and additional features. In the past (GPT 3.5 versus GPT 4.0), the difference in quality was quite noticeable. The paid version often provided better code and clearer descriptions, and it was much more up to date (due to training with newer material).

A major advantage of the Plus version is the ability to use specific GPTs. These are wizards that are optimized for specific tasks. For example, you can use the *Data Analyst* GPT to statistically evaluate and visualize data volumes. Simply drag and drop a file with the underlying data material into the chat history. The generated code gets executed immediately by ChatGPT, and you can view the result directly in the web browser.

You can also define your own GPTs with very little effort. An example of this follows in Section 1.2. Specific GPTs can significantly speed up the completion of some tasks.

Companies can activate ChatGPT Team for $25 per month per employee. The main difference compared to ChatGPT Plus is that the transmitted data isn't used for training ChatGPT. This is an advantage in terms of data protection, of course, but it doesn't change the fact that you still have to upload your code, your ideas and, ultimately, company secrets to the cloud to use ChatGPT. If you or your company have data protection concerns, you should consider running language models locally. We'll go into this topic in detail in the second part of the book (in particular, Chapter 9).

In our experiments, we've already had excellent experiences with the free ChatGPT version. The functions provided are great and absolutely sufficient for learning a programming language or for creating hobby applications.

But if you make a living from coding, $25 per month for more up-to-dateness and better-quality results is a good investment. However, you also pay for various additional features that aren't relevant for coding, such as imaging and audio/video functions. In this respect, the question arises as to whether a GitHub Copilot subscription (Section 1.3) might be the better investment.

Most of the ChatGPT results presented in this book were generated with version GPT-4o (4 *omni*). By the time you read this book, however, there will probably be newer versions of GPT that work even better.

1.1.7 Anthropic Claude

We've already referred to various ChatGPT alternatives (Google Gemini, Microsoft Copilot, etc.) in the introduction to this section. As far as software development is concerned, we became particularly fond of *Claude* from Anthropic (*https://claude.ai*) while we were working on this book.

Like ChatGPT, Claude can be used free of charge to a limited extent but requires a paid account for more intensive use. The web interface is very well designed and easy to use. In our tests, the quality of the results was at least on a par with those of ChatGPT, sometimes even better. As of December 2024, Claude was also characterized by a larger context window, meaning it could process more user data and longer listings. Just give it a try!

1.2 The Art of Prompting

The input or question in AI chat systems is usually referred to as a *prompt*. Asking the right questions has a major influence on how good or effective the results are. The art of optimally formulating the prompt is often referred to as *prompt engineering*. This section summarizes some tips on this topic:

- **Additional information**
  Introduce the question with contextual information about the programming language, library, tool, or environment you're using.

  If you ask for a regular pattern in general, ChatGPT has to guess which syntax variant is meant (Portable Operating System Interface [POSIX], POSIX Extended Regular Expressions [ERE], Perl Compatible Regular Expressions [PCRE], etc.). However, if you specify that you want to use the regular pattern in the C# language, ChatGPT can take the Microsoft-specific features of regular expressions into account.

  It's sufficient to briefly prefix the first prompt of a session with the programming language or another specification, such as "bash/Linux", "nodejs 20", "PHP 8", "Regex/PCRE", "SQL Server", or "zsh/macOS".

- **Additional conditions**
  Indicate what is important to you:
  - Do you want short, compact code?
  - Should as few imports/modules/libraries as possible be used?
  - Is the code security-critical?
  - Is maximum efficiency in execution crucial?
  - Should the code be formulated recursively, if possible? Or, on the contrary, should recursion be avoided?

  You can always dispense with such additional rules in the first attempt and wait and see what the result looks like. Afterward, you can still ask the AI tool to reformulate the code in compliance with new conditions: *The code works, but is too slow. Is it possible to make it faster?*

- **Formal rules**
  You can define formal rules. These can relate to the maximum line length, the desired convention for variable and function names (camel case, snake case, etc.), the location and extent of comments, and so on.

- **Language**
  Even if English is not your mother language, try to formulate your question in English.

- **Follow-up questions**
  Ask follow-up questions. ChatGPT automatically takes the question/answer history into account.

- **A second try never hurts**
  If you're not satisfied with an answer, just try again. Clearly state which detail of the answer doesn't fit from your point of view.

  In the ChatGPT web interface, you can click on the **Regenerate** button. The new answer happens to be a little different and may be more in line with your expectations. But as long as ChatGPT doesn't know why you're dissatisfied, the chances of a better result on the second attempt are relatively low. It's better if you state in a follow-up question what it was you didn't like in the previous answer.

  If there is no **Regenerate** button and you don't have the time or desire to formulate specific improvement requests, you can try the following universal instruction: Do better! Don't worry about the simplistic nature of this prompt; it requires little effort and occasionally even leads to success.

- **Multiple options**
  Usually, you ask a question and get *one* answer, but it may not be the best answer from your particular perspective.

  If you're familiarizing yourself with a new area or the solution is unclear, you should also express this in the prompt. Ask the language model to give you several options to choose from, for example:

  What options do I have to implement a REST API in Python?

  The AI tool will present an overview of various libraries and frameworks. You can research details of the proposals on the internet or in the AI chat system. Then, decide on one of the variants and ask follow-up questions.

- **Train of thought (reasoning)**
  The following prompt idea goes in a similar direction. Instead of requesting *one* specific solution, you can ask the AI chatbot to put itself in the role of several experts discussing a question like the following example:

  Imagine three IT experts working on a question. They discuss various proposals, discard unsuitable ideas, and finally agree on a solution. The task is as follows: How can a redundant backup system be designed for a web server?

- **Working step by step**
  For complex tasks, you should avoid trying to fit everything into one prompt at once. It works much better if you proceed step-by-step and first ask the AI chatbot for an outline and then take care of the details in further prompts. This procedure is also known as *prompt chaining*. The first prompt may look as follows:

I need to design a database to analyze our web traffic. First of all, please give me an outline of the database. (What tables do I need?) We'll discuss the details of each table later.

Note that ChatGPT and similar services have strict limits for the maximum input and output length (size of the context window and maximum number of output tokens). AI chatbots will therefore never provide answers that consist of several thousand lines of code. It's also not advisable to pass a file with many lines of code to the prompt and request a complete revision (e.g., refactoring). Instead, you need to break your task down into parts and carry out each step separately.

- **New session on change of topic**
  Start a new session when you change the topic. ChatGPT takes the information collected in a session into account for follow-up questions. All previous questions and answers therefore form the context for all further questions.

  However, if a new question relates to a different programming language, a different problem or a different library, it's advisable to start again in a separate session.

  A restart can also be useful if you're dissatisfied with the results so far or if you think that you or the AI tool have reached a dead end. The context from the preceding questions and answers may prevent ChatGPT from pursuing other ideas.

- **Understanding and testing code**
  The AI tool isn't responsible for your code—you are! Don't just hope that the generated code will work; instead, test it thoroughly, taking all conceivable special cases into account. If you can't understand the AI-generated code, you need to familiarize yourself with the underlying concepts, even if it's inconvenient at the moment.

- **Take particular care with new or exotic topics**
  As a matter of principle, language models have difficulty with new technologies or peripheral IT areas. On the one hand, there is little training material available; on the other hand, the problem is exacerbated by the fact that the training for a current language model started usually at least 6 months and often 12 or more months previously.

Colloquial Prompts Also Work, But Not Always Perfectly

In this book, we've endeavored to formulate the prompts in neat, complete sentences. The entire text of this book has been proofread, and prompts in poor English simply don't fit in.

But don't worry: AI tools usually understand you even if you formulate the prompt very briefly. Prompts such as the aforementioned *Do better!, Optimize this code!, How does this work?*, or *Why doesn't this work?* are also fine.

However, it's important that the intention of your question is unambiguously clear. A precise formulation helps both you and the AI tool better understand the problem. In this respect, the time spent on a "proper" formulation of a prompt is usually well invested.

1.2.1   Prompt Frameworks

Compliance with the rules of various prompt frameworks can help you with prompting. In this context, a "framework" is a kind of guide to the components from which a prompt can be assembled.

The most popular framework is the RTF framework. In the prompt, you formulate the *role* of the AI tool, the specific *task*, and the desired *format* of the response. In the following example, we've labeled the components of the prompt. In real life, of course, these markings are omitted.

> **Prompt:** (*Role*) You're an experienced Python developer. (*Task*) Sketch the framework of a REST API for a to-do app using Django. (*Format*) Present the answer in a step-by-step manner. Just show the outline of the code; don't go into details yet.

Alternatives to RTF are the RODES framework (role, objective, details, example, sense check) and the RISEN framework (role, instructions, steps, end goal, narrowing). However, considerable effort is required in formulating that many details. In this respect, the use of such frameworks is more suitable if you program an AI chatbot or another application yourself using an API (see also Chapter 10).

The use of the frameworks outlined here for the prompt structure is optional. Even if you only specify the task (i.e., the *task* in the sense of the RTF framework), you'll achieve useful results. But especially in tricky cases, adhering to framework rules can lead to better results.

1.2.2   A Question of Context

We've already pointed out that AI tools require additional information, that is, context, to provide relevant answers. The context determines which data is actively processed outside the actual language model. In a ChatGPT session, the context results from all questions (prompts) asked in a session so far and the resulting answers.

The size of the context of language models is limited. It's normally measured in tokens (see Table 1.1). More technical details about context and tokens will follow in Section 1.5. Some language models have further limits. The output of ChatGPT-4o is limited to 4,000 tokens. This means that ChatGPT-4o can only deliver approximately 12 KB of code at a time despite a comparatively large context memory. The output limit of

4,000 tokens also applies to the *Claude 3 Opus* language model from Anthropic. It was increased to 8,000 tokens for *Claude 3.5 Sonnet*.

Language Model	Max. Context Size	Max. Text Quantity	Max. Code Quantity
GPT-3.5 Turbo	4,096 tokens	~16,000 characters	~12 KB
GPT-4	8,000 tokens	~32,000 characters	~24 KB
GPT-4o (omni)	32,000 tokens	~128,000 characters	~96 KB
Llama 3.3	128,000 tokens	~500,000 characters	~380 KB
Claude 3 Opus	200,000 tokens	~800,000 characters	~600 KB
Claude 3.5 Sonnet	200,000 tokens	~800,000 characters	~600 KB

Table 1.1 Maximum Context Size of Various Language Models from OpenAI, Meta, and Anthropic (as of January 2025)

In the future, even the considerable context size of the Anthropic models could look modest. In August 2024, the Magic AI company presented a language model that can cope with a context window of 100 million tokens. This corresponds to approximately 10 million lines of code. For comparison, the entire kernel of Linux, which is one of the largest public software projects, comprises around 30 million lines of code.

What do the maximum context size (often also the *context window*) and the maximum output have to do with the prompt? We've had the best experiences with the use of AI tools when we've limited ourselves to small, manageable functionalities. Don't ask for everything at once ("Write code for hotel booking software like booking.com."); instead, first design the structure of your program, and then apply AI tools to its individual functions.

GPTs Are Stateless

You're used to ChatGPT and other tools allowing you to refer to the questions and answers from the previous chat history in follow-up questions. This gives you the impression that ChatGPT is memorizing the conversation.

However, language models actually work *statelessly*; that is, they don't remember anything and regard every prompt as completely new. The impression of memory is only created by the programming of the chat interface. From the second prompt, the website not only transmits the new prompt to the voice system but also all previous questions and answers. These form the context for the new question and therefore also the "memory" for the chat history.

In Chapter 10, we describe how the programming of a minimal chat system works behind the scenes.

1.2.3 Custom GPTs

If you repeatedly specify certain rules, it may be worth setting up a customized *generative pretrained transformer* (GPT). (We'll explain the term GPT fully in Section 1.5.) With ChatGPT, defining your own GPTs requires a paid version of ChatGPT. In the ChatGPT web interface, first click on **Explore GPTs** and then on **Create**. In the GPT editor, go to the **Configure** dialog box. There, you give your own GPT a name, briefly describe the intended use, and then specify the instructions for the GPT.

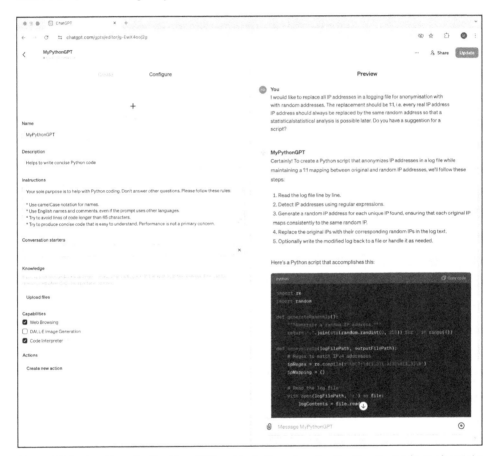

Figure 1.2 Instructions for the New GPT on the Left, and an Application Example on the Right

These instructions can look as follows, for example:

> **Instructions:** Your sole purpose is to help with Python coding. Don't answer other questions. Follow these rules:
> - Use English names and comments, even if the prompt uses other languages.
> - Use camelCase notation for names.
> - Try to avoid lines of code longer than 65 characters.

- Try to produce concise code that is easy to understand. Performance isn't a primary concern.
- If possible, try to avoid modules that need to be installed externally (`pip`).

Depending on the application, you can use the **Upload files** button to upload files with additional information (e.g., the documentation for a new software library or version that wasn't yet available when the language model was trained), which we haven't done in this example. Permitted document formats are plain text, Markdown, HTML, and PDF. You can try out your individual GPT right away in the **Preview** area.

In Figure 1.2, you can see that the new coding wizard works as intended:

- The prompt doesn't contain any information about which programming language ChatGPT should use. But due to the individual GPT instructions, ChatGPT knows that it should use Python.
- Function and variable names use camelCase notation.

The **Code Interpreter** checkbox allows the GPT to execute code directly, for example, to perform a test run. However, this only works for a few programming languages and is also associated with various restrictions, such as terms of runtime or access to files or the internet.

Prompt: Can you show me a script to calculate Fibonacci numbers? Please run the script and calculate the 25th Fibonacci number.

```
def fibonacci(n):
    """Calculate the n-th Fibonacci number."""
    a, b = 0, 1
    for _ in range(n):
        a, b = b, a + b
    return a

# Calculate the 25th Fibonacci number
fibonacci(25)
```

The 25th Fibonacci number is 75,025. If you need to calculate another Fibonacci number, you can use the `fibonacci` function provided in the script by changing the input value.

1.2.4 Comparing Custom GPTs, Fine-Tuning, and Retrieval-Augmented Generation

Custom GPTs primarily change the role of the language model, that is, its basic task. Usually, the role description is "You are a helpful wizard" or something similar. We'll go into the concept of role, prompt, and response in more detail in Chapter 10. A changed role influences the answers of the language model, but doesn't change its core.

You have even more influence on the functionality of a given language model via *fine-tuning*. In the simplest case, you give the language model a list of questions (prompts) and what you consider to be the best answers. The language model can be based on this. However, ChatGPT doesn't currently support fine-tuning for custom GPTs.

The "real" integration of additional knowledge that wasn't yet available when the language model was trained is provided by *retrieval-augmented generation* (RAG; see also Chapter 12). This procedure is also used for custom GPTs if you upload additional knowledge files.

Unfortunately, it's impossible to simply retrain a language model for a specific application or to take additional information into account. This approach consumes an extremely large amount of computing time and money and is therefore only feasible for large (AI) companies.

The Limits of Custom GPTs

Integrating a few PDF files with additional information and changing the system instruction (role) of the language model isn't enough to build a well-functioning company-specific or application-specific bot in just a few minutes. Such attempts have always failed miserably in the past.

1.3   Code Wizards

Chat-based AI tools provide an uncomplicated and free introduction to the world of AI-supported coding. However, the constant back and forth between the web interface of the AI tool and the editor or development environment is somewhat inconvenient. You constantly need to copy code back and forth between the editor and the web browser, and surely that must be possible in an easier way.

In fact, Microsoft had already started the technical preview of *GitHub Copilot* before OpenAI presented its universal ChatGPT interface to the public. GitHub Copilot has been available as a paid service since June 2022. The cheapest option currently costs $10 per month.

The "Copilot" Concept

Microsoft initially only used Copilot to describe the new code wizard. However, the term is now used widely for various Microsoft AI tools, for search functions, as well as for Microsoft Office tools. Since mid-2024, there has even been a notebook series with this name (Copilot+ PC). This book focuses exclusively on the GitHub Copilot.

The idea of GitHub Copilot is a logical extension of the IntelliSense concept that has long been established in the Microsoft world. While entering code, Visual Studio Code (VS Code) suggests a completion for the entry. But while IntelliSense was limited to the names of methods or properties, GitHub Copilot makes suggestions for loops, including their content, or even suggestions for entire methods or functions.

GitHub Copilot sometimes works incredibly well. Say you start the definition of a new function such as `findCancelledCustomers`. From the context of the rest of the file, GitHub Copilot recognizes that you probably want to get a list of all customers from the `customers` table whose status is **Canceled**. Accordingly, GitHub Copilot proposes the complete code for the function: it generates SQL code for the query, executes using the variable already defined elsewhere in the file for the database connection, and returns the list of customer names and IDs as the function result.

Of course, GitHub Copilot's suggestions aren't always perfect. You can often use the code anyway and then have to adjust a few details. If the AI tool doesn't correctly guess your intention, simply continue entering your code manually. In the subsequent line or the line after that, GitHub Copilot will then make new suggestions that are better due to additional context information (e.g., as comments or variable names).

Behind the scenes, GitHub Copilot, like ChatGPT, currently uses a language model from OpenAI (most recently GPT-4). The language model was trained using code from GitHub's public repositories, among other things. This fact has led to controversies concerning both copyright issues and security aspects. It's still not entirely clear whether training with code from public sources is actually considered *fair use*, as argued by GitHub and Microsoft. In this book, we address this question and other aspects in Chapter 13.

Alternatives to GitHub Copilot

Just as ChatGPT isn't the only chat-based AI environment, there are also various alternatives to GitHub Copilot, such as GitLab Code Suggestions with *Duo Chat* or the *Swift Assist* function integrated in current Xcode versions.

Cody is extremely exciting. Basically, Cody also completes code and answers chat prompts. But it has two main advantages: First, you can choose between different language models (local large language models [LLMs], Claude Sonnet, Gemini, Mixtral, and GPT-4o and Cloud Opus with the Pro account). Second, Cody takes into account your entire code base in the active directory. (This function is expected to also be implemented in GitHub Copilot sooner or later.) You can try Cody free of charge. The Pro subscription with unlimited chat functions and a larger selection of LLMs currently costs $9 per year. You can find more information on this here: *https://sourcegraph.com/cody*.

Apart from commercial offers, there are also free editor extensions and plug-ins. They require an AI system to be accessible via an interface (API). In this book, we'll focus on the VS Code plug-in *Continue* as an example (see Chapter 9, Section 9.5). It provides very similar functions to GitHub Copilot, but can use a locally executed language model. Provided you have the appropriate computing power, you can avoid having to pay for AI providers and avoid transferring your data to the cloud.

Continue can also be connected to an AI instance in the local network or on the internet. For example, you can configure a powerful computer with a GPU as a workgroup; all team members then use this computer together for AI coding tasks. A major advantage for companies is that they can avoid any data protection concerns.

1.3.1   Installing GitHub Copilot

GitHub Copilot was initially developed as an extension to the VS Code editor. The combination of VS Code with GitHub Copilot is still extremely popular today. Accordingly, most of the code wizard examples in this book are applications of this "dream team." Before you can use GitHub Copilot, you need to install the extension of the same name, log in to your GitHub account in VS Code, and sign up for a Copilot subscription on the GitHub website (*https://github.com/features/copilot/plans*).

You can try GitHub Copilot free of charge for 30 days. Free licenses are provided for students, teachers, and maintainers of large open-source projects.

Editors Compatible with GitHub Copilot and Development Environments

GitHub Copilot can be used not only with VS Code but also with various other editors or development environments, such as Visual Studio, all JetBrains IDEs (IntelliJ, Android Studio, PyCharm, etc.) as well as the Vim and Neovim editors. There are unofficial plug-ins for various other editors, such as Emacs or Xcode.

1.3.2   First Steps with Visual Studio Code and GitHub Copilot

After the installation, VS Code displays a Copilot icon in the status bar that leads you to the Copilot status menu. The first line of the menu must contain the **Status: Ready** entry.

To try out GitHub Copilot, create a new file in your favorite language. We've used Python for the following example. Now enter def factorize(x), for example. It can be assumed that you want to program a function that breaks down an integer into its factors.

GitHub Copilot suggests a possible continuation of the code in gray font. If you want to accept the entire proposal, simply confirm it by pressing the [Tab] key. Alternatively, you can repeatedly press [Ctrl]+[▶] to accept the suggestion in small parts (word by word or line by line). The mini dialog for proposal selection only displays when you move the cursor over the code proposal.

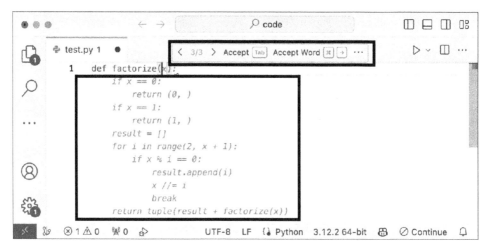

Figure 1.3 First Experiments with GitHub Copilot in VS Code

1.3.3   Switching between Code Suggestions

VS Code always shows only one code suggestion, although GitHub Copilot internally offers several options to choose from. The shortcut keys ⌨Alt+⌨[or ⌨Alt+⌨] are used to switch between the first three suggestions.

Keyboard Shortcut	Function
⌨Tab	Accepting a suggestion
⌨Ctrl+⌨▶ or ⌨cmd+⌨▶	Accepting a word or line of the suggestion
⌨Alt+⌨[	Next suggestion (US keyboard)
⌨Alt+⌨]	Previous suggestion (US keyboard)
⌨Ctrl+⌨Enter or ⌨cmd+⌨Enter	Displaying 10 suggestions in a side pane
⌨Ctrl+⌨I or ⌨cmd+⌨I	Opening the Copilot dialog (inline chat)
⌨Ctrl+⌨Alt+⌨I	Opening the chat sidebar
⌨Ctrl+⌨cmd+⌨I	Opening the chat sidebar (macOS)

Table 1.2 VS Code Shortcuts for Controlling GitHub Copilot

These shortcuts aren't defined on Linux and macOS, but you can change that by means of a manual configuration.

We opened the **Keyboard Shortcuts** configuration dialog in VS Code, searched for the editor.action.inlineSuggest.showPrevious and .showNext commands, and assigned our own keyboard shortcuts to them by double-clicking in the **Keybinding** column.

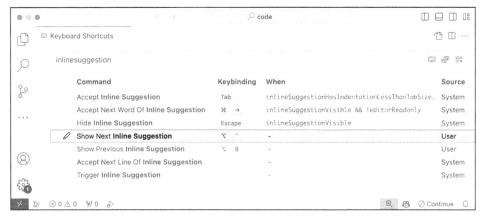

Figure 1.4 Defining Custom Shortcuts for Selecting GitHub Copilot Suggestions

If three suggestions aren't enough, you can press Ctrl + Enter to open a pane on the right-hand side with up to 10 suggestions. You can let yourself be inspired by the ideas and then select a suggestion by clicking on the **Accept suggestion n** button.

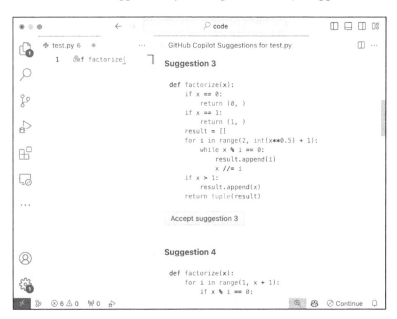

Figure 1.5 Code Suggestions in the Pane on the Right

1.3.4   Tips and Tricks

When testing in a new, almost empty code file, a significant part of the potential of GitHub Copilot remains unused. We therefore recommend that you familiarize yourself with the AI tool in a project whose code you know well. It's crucial that you can assess the quality of GitHub Copilot's suggestions beyond doubt. This works best in a familiar environment. Just add a new feature to your project or make changes to exist-ing code, and you'll be amazed at how constructive and useful the suggestions are!

To exploit the full potential of GitHub Copilot, consider the following tips:

- Give new functions or methods names that are as meaningful as possible.

- If necessary, formulate a comment before the function definition, before a loop, or before a branch (one or two lines are sufficient) to describe the task more precisely or to point out special features or details. This will help GitHub Copilot make more appropriate suggestions in context.

- You can also take the opposite approach by writing code first (with or without AI support) and then letting GitHub Copilot help you write the comment. Simply start the comment with #, //, or /*, depending on the language. GitHub Copilot parses the surrounding code and suggests documentation that is often useful.

- Open a few relevant other files of your project in VS Code with comparable code or with the definition of functions or classes that you want to use in the file you're cur-rently working on. GitHub Copilot considers the code of the open tabs as context information and recognizes how you've called or evaluated functions, properties, or methods so far. This improves the quality of the code proposals.

  Currently (early 2025), GitHub Copilot only considers project files that are open in the editor. It's possible that future versions will also analyze closed files. This is cur-rently not working for two reasons: First, the size of the context is very limited with the current language models and GitHub Copilot simply can't take all files into account. Secondly, with large projects, it's difficult to identify the relevant files within hundreds of files.

1.3.5   Chat Mode with Additional Keywords

The code wizard and the chat functions are by no means mutually exclusive. Press Ctrl+Alt+I or ctrl+cmd+I to open the chat sidebar within the editor. There you can communicate freely with the AI language model. You can transfer the code contained in the response directly to the currently open code file by clicking a button at the current cursor position, thus saving yourself the detour via the clipboard.

A decisive advantage compared to ChatGPT in the web browser is that you can refer to the context of the open editor tabs when communicating via special keywords, as fol-lows:

- @workspace indicates that you're referring to the code in the workspace and don't want to ask a general question. (The VS Code workspace is made up of the files in the project directory or all currently active projects.)

- @vscode means that you're asking a question about using the editor. GitHub Copilot may include VS Code commands in the response that you can click on directly.

- `@terminal` directs the context to the terminal, which is useful for questions relating to terminal commands. Depending on the operating system used, GitHub Copilot explains answers to Linux and macOS commands or PowerShell cmdlets.

- `/clear` deletes the preceding chat history.

- `/doc` adds comments to the code.

- `/explain` explains the task or function of the current code. This is particularly useful if you're editing third-party code or have taken code from the internet (e.g., from Stack Overflow) and don't fully understand it.

- `/fix` tries to fix problems or errors in the selected code.

- `/tests` generates unit tests for the current function or method.

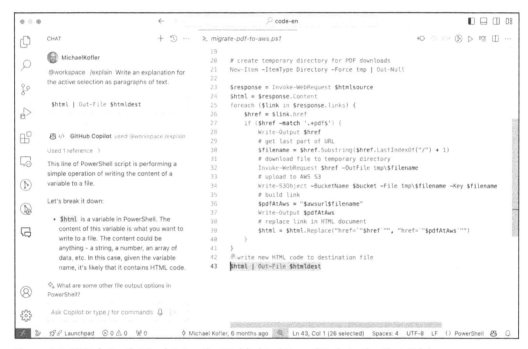

Figure 1.6 GitHub Copilot Explaining the Highlighted Line of Code in the Chat Sidebar

Tip

The `/explain`, `/fix`, `/doc`, and `/tests` commands can be run even more conveniently directly in the code window. To do this, select the relevant lines in the code, and execute one of the following context menus: **Copilot • Explain This, Fix This, Generate Docs**, or **Generate Tests**.

Documentation and unit tests are covered in detail in Chapter 6 and Chapter 5, respectively.

1.3.6   Chat Examples

The following prompts show some examples of how the keywords are used. In the first example, VS Code is supposed to explain the code just highlighted.

Prompt: @workspace /explain the selected lines.

The selected lines are from a shell script and perform the following operations . . .

You can use GitHub Copilot to search for specific files in the project. Our tests produced mixed results. The CSS search was successful, while the search for unsafe full-text passwords failed.

Prompt: @workspace Which CSS file defines the invoiceDate class?

The invoiceDate class is defined in the css/invoice.css file.

Prompt: @workspace Which scripts contain unsafe clear text passwords for database connections?

[no useful results]

If you're lost in the plethora of VS Code commands, GitHub Copilot can sometimes help you out. Again, here's a positive and a negative example. The answer to the second prompt is incomplete because GitHub Copilot doesn't provide the option of activating the secondary panel and simply dragging the chat there (see also the corresponding "Making the Best Use of Space" box).

Prompt: @vscode How can I see the differences in the code file since the last commit?

You can use the **Git: Open Changes** (git.timeline.openDiff) command to show the differences in the active document compared to the last commit.

Prompt: @vscode I want to display the chat panel on the right side. All other panels should remain at the left side.

You can use the **View: Move Panel Right** command to move the panel to the right. Then, use the **View: Show Chat** command to open the chat panel.

Please note that this will move all panels, including the chat panel, to the right side. VS Code currently doesn't support displaying panels on different sides simultaneously.

When executing the following question, we've just edited files on a Linux server in VS Code. We particularly liked the correct one-line answer without any superfluous blah-blah.

> **Prompt:** @terminal How can I find files that haven't changed for more than six months?

```
find . -type f -mtime +180
```

You can't directly change chat commands once they have been executed. Instead, you need to scroll through previously executed prompts using the cursor keys as in the terminal, change them, and then execute them again.

Making the Best Use of Space

If the chat in the VS Code sidebar is too small for you, you can open it in a separate window via **Open Chat in Editor** or **Open Chat in Window**.

Another variant is to use **View: Toggle Secondary Side Bar Visibility** to show a second sidebar on the right-hand side and move the chat history there. This is particularly useful when you work on very wide monitors.

1.3.7   Alternatives to the Chat Sidebar

Instead of using GitHub Copilot in the chat sidebar, you can open a small inline chat dialog via `Ctrl`+`I` or `cmd`+`I`, respectively. You can formulate your question there and then transfer the answer directly into the code. The inline chat can be used very efficiently for small changes. In addition, the chat history remains organized.

You can also ask minor questions directly in the code. To do this, start a comment using q:, and then formulate your question. As soon as you press `Enter`, GitHub Copilot inserts another comment line that starts with a: and contains the answer:

```
// q: How do I call bash script from PHP?    <- Your input
// a: Use shell_exec() or exec() function.   <- GitHub Copilot
```

This type of question/answer game is particularly suitable for small details. The biggest advantage is that you never have to leave the editor. As soon as the problem has been solved, you can delete the comment lines containing the question and answer.

```
80    // write to /tmp/accountdata, call bash script /etc/myscripts/makeaccount
81    function setupNewAccount($product, $customer) {
82      $myfile = fopen("/tmp/accountdata", "w");
83      fwrite($myfile, 'USER1LASTNAME="' . $customer->lastname . '"' . "\n");
84      fwrite($myfile, 'USER1FIRSTNAME="' . $customer->firstname . '"' . "\n");
85      fwrite($myfile, 'USER1EMAIL="' .    $customer->mail . '"' . "\n");
86      fwrite($myfile, 'ACCOUNT="' .       $customer->url . '"' . "\n");
87      fwrite($myfile, 'DB="' .            $customer->db . '"' . "\n");
88      fwrite($myfile, 'MAXUPLOAD="' .     $product->storage . '"' . "\n");
89      fwrite($myfile, 'LANG="' .          $customer->lang . '"' . "\n");
90      fclose($myfile);
91
```

I want to run a bash script with sudo here. Can you help? Don't remove existing code!

GitHub Copilot

Made changes.

Accept   Discard   Accept or Discard 1 change.

```
92    // Run the bash script with sudo
93    $command = "sudo /etc/myscripts/makeaccount";
94    exec($command, $output, $returnCode);
95    if ($returnCode !== 0) {
96      echo "<p>Error running the bash script: $returnCode";
97      exit;
98    }
99  }
```

Figure 1.7 The Inline Chat Dialog

1.3.8   Other Functions

Microsoft is endeavoring to integrate GitHub Copilot as an all-encompassing tool in VS Code. You must decide for yourself to what extent these functions are useful in your workflow:

- **Git commits**
  If you use Git version management, you can have the commit text written by GitHub Copilot. The AI tool attempts to summarize the last changes made. Sometimes, this works quite well, and, at other times, you just need to change the text. To us, the messages tended to be too lengthy.

- **Process error message**
  If an error occurs during the execution of a program, VS Code displays a **Copilot** icon in the terminal area next to the error message. If you click on it, GitHub Copilot will try to explain the cause of the error in the chat. Ideally, it also provides a solution straight away and executes the suggestion with a further click.

  Our experience with this feature was mixed. For some programming languages (including Python!), GitHub Copilot doesn't even recognize completely trivial errors such as missing parentheses. Admittedly, there are also cases in which the function works as advertised. But there is certainly still room for improvement here.

- **GitHub Copilot in the terminal**

  Once you've executed a command in the terminal, you can use GitHub Copilot to have this command explained to you.

  If an error occurred during the execution of the last command, GitHub Copilot explains the cause of the error and suggests a corrected or improved command.

  We've already pointed out the possibility of asking questions in the chat with `@terminal` that relate specifically to terminal commands.

1.3.9   Security and Data Protection

You should be aware that GitHub Copilot has access to all files in your project. This is necessary for the tool to work. The access also includes files with passwords, license keys, and so on. You must therefore take a great leap of faith with GitHub Copilot that it won't misuse this data.

Ultimately, however, this consideration applies not only to AI tools but also to every editor and every development environment: a backdoor in the IDE or a plug-in with malicious code can also leak company secrets. You should therefore think carefully about which plug-ins you install in VS Code or the IDE of your choice. It's actually surprising that there has never been a (known) safety disaster in this regard.

Never Save Passwords in Code Files!

Irrespective of the preceding considerations, you should never save passwords directly in code files. Use separate configuration files such as `config.json` or `.env`, or save passwords in environment variables that your code can access.

Of course, VS Code also has access to password files. In most cases, however, their evaluation by GitHub Copilot or other AI tools isn't necessary. However, if a password is in plain text in the code file, it will inevitably be uploaded to external servers when AI tools are used. Even if the AI providers themselves have no interest in these passwords, their uncontrolled storage in any database to improve or train future language models is a security nightmare.

1.3.10   Trouble in Remote Mode

The *Remote SSH* plug-in allows you to use VS Code to also edit code in directories located on other computers, provided you have Secure Shell (SSH) access on the external computers. In our tests, however, GitHub Copilot refused to cooperate in such cases. The following lines in the `settings.json` configuration file on your local computer can help:

```
# include in settings.json
...
"remote.extensionKind": {
    "GitHub.copilot": ["ui"],
  },
...
```

The easiest way to open the `settings.json` file is by using **Open User Settings (JSON)** (F1). You can find more details on that process here: *https://github.com/orgs/community/discussions/6942#discussioncomment-4962858*.

1.4   Chat or Wizard?

You may be asking yourself whether you need a paid AI coding subscription, or whether free web services such as ChatGPT and Anthropic Claude are sufficient. First of all, the question is incomplete in this form: provided you have a powerful computer, you can also implement the coding wizard functions with local language models, that is, without a subscription. We'll go into this topic in more detail in Chapter 9. This avoids both the subscription and data protection concerns. However, costs are still incurred—either for you privately or for your company: you need an expensive computer, which in turn needs more power, and so on.

But regardless of the details of how you implement your code wizard, the basic question still remains: Would you prefer to work chat-based or follow the ideas of a wizard without being asked while you're entering code? In our everyday lives, a combination of both techniques has become the established approach. Now that we've become accustomed to the convenience of a coding wizard, we no longer want to do without these functions. Nevertheless, we often find it useful to research basic questions (less code details, but rather strategic considerations) in a chat window or web browser, separated from the editor or the development environment.

But the question of whether you need a chat or assistant also has to do with personal preferences. Of course, it's more complicated to find out details in the chat first and then implement them manually in the code. Frequent copying and pasting is a major source of errors, for example, if variable and function names suddenly no longer match. Nevertheless, this workflow gives you better control over your code. Ultimately, you're responsible for your code and must understand it.

> **Further Reading**
>
> You can read an extremely exciting interview about various applications of AI tools in the "Building a Text Editor in the Times of AI" article here: *https://zed.dev/blog/building-a-text-editor-in-times-of-ai*.

In the interview, the developers of the Zed editor describe how they use AI tools themselves and how they integrate AI functions into the new editor.

1.5   Basic Principles of Large Language Models

To use AI tools in coding, you don't need to understand how they work. In this respect, you can simply skip this section. Nevertheless, it's useful if you at least know the most important terms relating to *large language models* (LLMs).

> **More Depth**
>
> Don't worry, this chapter isn't a comprehensive AI basics course! There are already enough books available on this subject. We can particularly recommend *Developing AI Applications: An Introduction* by Metin Karatas, also published by Rheinwerk Computing.
>
> If you want to get to know and understand AI and LLM basics with more technical details, we also recommend Mario Zechner's basic course. The course consists of multiple Jupyter notebooks that invite you to experiment: *https://github.com/badlogic/genai-workshop*.

1.5.1   From the Human Brain to Neural Networks

Let's get started with the human brain, which consists of around 85 billion, that is, approximately 10^{11}, nerve cells (neurons). Each neuron is connected to many other neurons by synapses. In total, there are approximately 100 trillion = 10^{14} such connections. During learning, individual synapses are strengthened or weakened. So, when you memorize something or when you learn something, the synapses of your brain (i.e., the "wiring") change.

Compared to a computer, the human brain is amazingly good at processing information in parallel: seeing, hearing, thinking, and speaking. The brain also uses energy extremely efficiently and is content with around 20 watts. That is about one-fifth of the energy expenditure of an adult and significantly less than is needed to run LLMs on a computer.

For around 80 years, attempts have been made to recreate neurons and synapses as much as possible using artificial neural networks. From a mathematical perspective, a neural network receives and processes digital data (input), and then returns the result (output). Input and output are represented as vectors, that is, as sets of numbers. The content of the vectors depends on the application: text, audio data, the pixels of an image sensor, and so on.

The input vectors are processed into an output vector by matrix multiplication. This process usually involves multiple stages. The numbers ("weights") contained in the matrixes are also referred to as the parameters of the neural network. They correspond to the synapses of the human brain.

The first neural networks failed for two reasons: the number of parameters was too small to create an adequate reflection of reality, and the computing power was too low to train the model sufficiently. Today's language models consist of many billions of parameters. graphics processing units (GPUs) help to train the model. Increasingly sophisticated algorithms for designing the mathematical model make this concept extremely successful today.

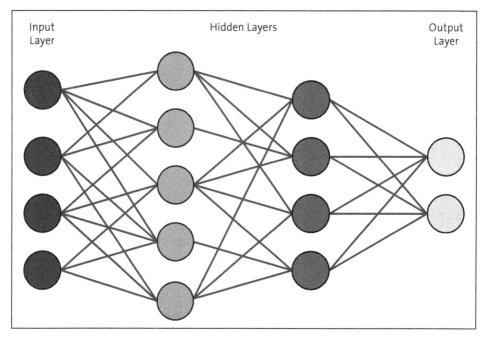

Figure 1.8 Diagram of a Neural Network

1.5.2 Machine Learning through Backpropagation

Self-learning algorithms for machine learning were an important milestone in the development of neural networks. One learning concept is to train the model with large amounts of familiar learning material, such as images of plants. With each training image, the contents of the matrixes are changed via *backpropagation* so that the deviations between the known results of the training material and the results calculated by the neural network are minimized. The program is then able to recognize the plants depicted in new photos (which were not part of the training) with a high probability.

Training can be implemented in an even smarter way by using systems that are clearly defined by rules: A computer program can play chess against itself and optimize the matrices of its neural model so that it wins as many games as possible. Today, computer programs based on the AlphaZero algorithm play chess and Go better than any human—and that without access to opening libraries and without practice material from world champion games.

Current AI systems are optimized for a specific purpose. They can't be used universally for every conceivable task. The optimization concerns both the structure and linking of the neural network as well as the learning algorithm.

Compared to the human brain, neural networks are still "simple." There are many more connections with different neurotransmitters in the human brain. The overall structure is therefore finer meshed and more varied than the artificial replica. In addition, the human brain can reorganize itself to a limited extent.

For all of these reasons, the human brain is superior to its artificial counterparts in its universal orientation. On the other hand, a computer (with or without neural networks) can perform special tasks faster.

1.5.3   Large Language Models

LLMs are a special case of neural networks that are specifically optimized for dealing with language. LLMs "understand" text instructions and can answer questions. Programming languages are ultimately just languages, albeit quite specialized ones. LLMs that have been trained with a sufficient number of code examples can therefore program surprisingly well. All the AI tools described in this book are based on LLMs.

Children begin to learn a language by hearing parents and other people speak, articulating their first words often still quite indistinctly. In nursery school, they sing songs with other children. They learn to read and write at elementary school. Over the course of nearly 20 years, the child acquires more and more words and the underlying grammar. They constantly receive feedback, interact with their environment, and learn how to communicate as well as possible. The teenager can now communicate by speaking and writing, understands complex texts, and can formulate them themselves. If you're reading this book, you've already largely completed this learning process. Nevertheless, you're still expanding your vocabulary with new technical terms!

LLMs work in a very similar way. During training, which only takes a few months in enormous server farms, the neural network is fed with gigantic amounts of text—the entire Wikipedia, the code collected on GitHub, entire libraries of books, Facebook postings, and so on. (The extent to which this training is permissible at all without the consent of the authors of these texts must first be clarified in court proceedings.)

What Training Material Was Used?

Presumably because of the legal uncertainty just mentioned, AI companies are revealing less and less about which data they have used for training. The sources of the free language model Llama 1 are broken down relatively precisely. The training material consists of approximately 1,200 billion tokens (*https://huggingface.co/datasets/togethercomputer/RedPajama-Data-1T*). This corresponds to around 2.4 billion pages of text in the format of this book, or almost 7 million books with roughly the same amount of content as this book.

Almost 90% of the training material comes from various publicly accessible websites, around 5% from GitHub and around 2% from Wikipedia. The rest of the materials are publicly accessible scientific publications (arXiv) and Stack Exchange contributions.

Various training data sets are available to download from the internet for AI research: *https://huggingface.co/datasets*.

Current LLMs store this knowledge in models comprising many billions of parameters. In principle, these parameters correspond to the weighting of connections between the nodes of the neural network. (In the human brain, too, the synapses between the neurons are designed differently; that is, they transmit a signal more or less strongly. During learning, the synapses change in the brain, and the weightings change in the neural model.)

LLMs differ not only in the number of parameters but also in the accuracy (*quantization*) with which they are stored. Often only 4 bits are used for this. This means that only 16 different values can be saved, but two parameters can be stored in 1 byte to save space.

For example, the freely available language model llama3:8b comprises 8 billion parameters (8b in llama3:8b stands for 8 billion). The total memory requirement of this model is around 5 GB because in addition to the parameters, some other data is stored.

There are currently attempts to save parameters with only 1 bit or with 1.58 bits (b1.58). The odd number results when *three* states are saved for each parameter: 0, 1, and -1. This kind of economical quantification allows for models with even more parameters with the same amount of memory. However, it still needs to be proved whether this is really advantageous. The efficient use of these so-called 1-bit LLMs also requires new hardware.

Bigger Is Better or Small Is Beautiful?

Language models with more parameters understand more complex questions and usually give better answers. However, they also require more resources, both during training and later in their application. The language model of ChatGPT-4, which isn't publicly accessible, probably comprises more than 1,000 billion parameters. The largest variant of the free Llama language model has 400 billion parameters (as of mid-2024).

However, the number of parameters isn't the only criteria. The quality of language models is heavily dependent on the material used during training and in fine-tuning. Finally, language models can be focused on content. Some LLMs are explicitly intended for programming. For this reason, a particularly large number of code examples were used during the training.

One of the most exciting AI research topics at the moment is the attempt to make high-quality language models as small as possible. *Small language models* (SLMs) would allow local application on notebooks and even smartphones. "Small" is relative; such models still comprise several billion parameters.

One approach isn't to use terabytes of undifferentiated texts collected from the internet for training, but to select less, higher-quality training material. It remains to be seen whether and when the quality of such "small" models can keep up with large commercial models.

1.5.4   The Transformer Model

A special feature of LLMs is the underlying transformer model. The model is based on a research breakthrough published in 2017 under the title "Attention Is All You Need" (available free at *https://arxiv.org/abs/1706.03762*). This relatively new AI technique applies neural networks to sequences, for example, to sentences made up of words. In the case of LLMs, however, *tokens* rather than words are used as data units: these are parts of a word, comparable to partial words or syllables.

Tokens are mapped internally by vectors. A vector is simply a sequence of numbers (an array). These vectors are relatively large and consist of 768 or 2,048 values, for example. Each value is mapped as a 16-bit or 32-bit floating point number. The space requirement of a token is therefore several kilobytes. The same word would only take up a few bytes in UTF-8 encoding.

The lavish use of internal representation has the advantage that a token doesn't simply store the characters of the word, but countless pieces of contextual information. Words that are related in content have similar vectors. The *tokenizer*, which is upstream of the transformer, is responsible for converting words into token vectors.

In general, a transformer turns a data sequence into a new data sequence. An easy-to-understand application is the translation of a text from one language into another. An encoder converts the input into an internal display. After processing, the decoder takes care of turning this data back into text.

The transformer model is characterized by the following features:

- The self-attention mechanism evaluates a word or its token in the context of one or more sentences.
- The transformer models process sequences in several successive layers (self-attention, feed-forward, etc.). This multilayer approach helps to recognize complex relationships.

- The transformer model dispenses with feedback (called *recurrent layers*). Compared to other AI models for language comprehension, this is more efficient. However, additional information about the position of words within a sentence must now be stored using *positional encoding* so that the logic of a sentence or that of computer code isn't lost.

During training, the transformer compresses the knowledge collected in huge amounts of text into comparatively small parameter sets. For example, 100 TB of text (10^{14} bytes) can be processed into a language model with 20 billion parameters. Its space requirement is 10^{10} bytes = 10 GB with a 4-bit accuracy per parameter.

Transformer Alternatives

The transformer model has helped LLMs achieve a breakthrough. It works amazingly well, but there are disadvantages. In particular, the resources required for training and application are very high.

AI research is therefore endeavoring to improve the transformer model or to take completely different paths to achieve equivalent results with less computing effort. It's possible that the transformer model will be a thing of the past in a few years' time, and new approaches will prevail.

If you're interested in AI research and are ready to dive into technical details, we have two reading tips for you:

- "xLSTM: Extended Long Short-Term Memory" (*https://arxiv.org/abs/2405.04517*)
- "Scalable MatMul-free Language Modeling" (*https://arxiv.org/abs/2406.02528*)

1.5.5   Generative Pretrained Transformer (GPT)

Chat Generative Pretrained Transformer (ChatGPT) and comparable programs can communicate in text form and are chatbots. The special feature compared to conventional, less sophisticated chatbots is that a previously trained transformer language model is used.

The central term *generative* in ChatGPT refers to generating new text to answer a question (a "prompt"). GPTs differ in various details from the original transformer model, which consists of an encoder and a decoder. The most important innovation is that GPTs focus on the decoder aspect: the decoder-only architecture doesn't convert existing text into another form as before, but creates new text to answer the given prompt.

An autoregressive model is used to generate the response: once a sentence has been started, it decides which is the most likely or most appropriate continuation in the given context. In this way, the answer is produced word for word without there being an internal representation of the entire answer beforehand.

Due to a certain randomness in the decision-making process (e.g., to choose between two similarly matching sequels), each answer is individual. This is why GPTs always provide varying answers to the same question. If you're not satisfied with the first answer, you can regenerate it with most programs and, with a bit of luck, get a better result.

Temperature

With some AI tools, you can set the "temperature." This option is available for almost all language models that you use via an API. The temperature determines how random the answer is. When AI tools are used for coding, too much randomness, that is, a high temperature, is rarely a good idea. Just think of the result if you're sick and feverish and try to finish programming something quickly.

On the other hand, there are cases where a little more creativity is required. Then, it may be worth experimenting with the AI temperature.

Even though the abbreviation GPT was coined as part of the ChatGPT brand name, GPT is now considered a generic term for LLMs that work in a similar way to ChatGPT.

1.5.6   Training: From Training Material to Language Model

The autoregressive model of GPTs works so well because the text prediction was previously practiced with millions of pages of different text. During training, the language model tries to guess the next word. The model is optimized step-by-step on the basis of deviations from existing training texts. Conveniently, this training can be carried out almost fully automatically.

Many billions of matrix multiplications must be performed during training. This is currently achieved most efficiently with GPUs or GPU-like processors. The largest manufacturer of such GPUs is NVIDIA. This is the reason for the huge current market value of this company.

In the course of the training, texts amounting to several terabytes are processed (1 terabyte of text corresponds to about 500 million book pages, which is about 1.7 million different books with a similar volume to this book). Huge server farms with GPUs are used for training.

Optimization of Training Material, Fine-Tuning, and Censoring

Further processing steps often take place before, during, or after training, which we won't discuss in detail here:

- During *data set curation*, unsuitable content (e.g., pornographic or racist texts, depending on the objective) is removed from the training material.

- The model is optimized during *fine-tuning*. For example, *biases* contained in the training material are corrected.
- *Censoring* attempts to prohibit the model from answering certain questions (e.g., "How do I produce a nerve gas?").

The result of the training is the finished language model, that is, the compact storage of the parameters that have just been calculated. The file of a language model is significantly smaller than the training material. Depending on the number of parameters, the size ranges from a few gigabytes for models intended for the local execution to several hundred gigabytes for the very large commercial models from OpenAI and others.

We often read that training is very energy-intensive and that huge amounts of training material are processed during training. Could this perhaps be a little more precise? To avoid being seen as energy wasters, AI companies are now more reticent about providing detailed information on the training process.

We've found relatively precise data on the free Llama 3.1 language model. In July 2024, Meta completed the training of the model, which is available in three sizes (8, 70, and 400 billion parameters). The training material consisted of approximately 60 TB of text. The training of all three variants required a total of almost 40 million GPU hours with an output of 700 W per GPU. This results in an energy requirement of just under 28 GWh. You can read more details about the model here:

https://github.com/meta-llama/llama-models/blob/main/models/llama3_1/MODEL_ CARD.md

It's hard to imagine how much 28 GWh of electricity is, so here's a comparison: Private electricity consumption per household in Germany amounted to an average of 3,400 kWh per year in 2021. Llama 3.1's training therefore consumed about as much electricity as 8,200 private households in an entire year.

1.5.7 Large Language Model Application: From Prompt to Response

Once a precalculated language model is in place, it's still a long way from sending the question (the prompt) to displaying the answer (the response). If you use ChatGPT, the result appears virtually in real time, or at least faster than you can read it. The high speed belies the number of processing steps required:

- **Tokenization/encoding**
  First, your input is converted into tokens that the language model can understand.
- **Token generation**
  An algorithm produces a sequence of output tokens based on the language model and the input tokens. Basically, each additional token is calculated as the most probable or plausible continuation of the previous output tokens.

- **Decoding**
The token stream is then converted back into language that humans can understand.

- **Postprocessing**
Various downstream processes ensure that the sentences are grammatically correct, that the punctuation is correct, and so on. The answer is also presented in an appealing layout (headings, highlighting, separate font for code, etc.).

These steps require billions of arithmetic operations. The effort becomes noticeable as soon as you try to run a language model locally on your computer (see Chapter 9). For slightly larger models with more than 10 billion parameters, even powerful notebooks with a lot of RAM are overwhelmed. The answers appear agonizingly slowly on the screen while the CPU or GPU are running hot. You're then faced with the following dilemma:

- You use an SLM. It's quite possible that the future belongs to task-specific language models. (A language model focused on coding doesn't necessarily have to shine with biological, geographical, or mathematical knowledge.)

- You purchase expensive hardware, for example, a desktop computer with a powerful graphics card. This significantly increases the speed, often measured in tokens per second. However, the model size is still limited by the memory (video RAM [VRAM]) accessible to the GPU.

- You stick to cloud-based solutions, which may be subject to a charge (as with GitHub Copilot).

LLM-Compatible Hardware

In the coming years, we expect notebooks to become LLM-capable, that is, capable of running language models locally with good quality and speed. This requires two things: sufficient RAM and hardware that can perform many matrix multiplications in a short time. The way there is either via more powerful GPUs (graphics processors) or via special *neural processing units* (NPUs).

NPUs are already integrated into modern CPUs, which promise a computing power of several teraflops per second (1,000 billion floating point operations per second). However, as of early 2025, it's fairly unclear what NPUs will actually be used for. Marketing is certainly the most important point. In second place is cropping out the person in front of the notebook camera in video conferences.

However, there is a lack of software drivers and established standards for the use of NPUs in the execution of local language models. In addition, the computing power of NPUs is currently significantly lower than that of GPUs. In this respect, it remains to be seen whether NPUs can really accelerate the execution of local language models.

1.5.8 The Limits of Linguistic "Intelligence"

Current language models can explain recursion in the language of Shakespeare. The result is fantastic!

> **Prompt:** Explain recursion to me in poem form. Try to imitate the language of Shakespeare.

Recursion, harken! A most curious art,
Wherein the selfsame path doth oft restart.
A function, bold, doth to itself return,
And through this loop, the mind begins to yearn.
[Three more stanzas follow, all in a similar quality.]

As amazing as the linguistic qualities are, you must never forget that you're dealing with a language model. Such models are subject to major restrictions in many respects. There is absolutely no indication of a "general intelligence," as evidenced here:

- Language models can't calculate.
- They can't draw (e.g., create diagrams).
- They only understand text, so they can't handle image, audio, and video data.
- They fail to reach logical conclusions that are relatively simple for humans.
- Their knowledge ends with the cut-off time for training. Language models can't access up-to-date information from the internet.

You may now object that these restrictions do not (all) apply. In fact, it looks as if modern AI chatbots are quite capable of calculating, evaluating current websites, following voice instructions, or describing images. However, all of this is only possible if language models are extended with external additional functions using various tricks. A few typical procedures are summarized here:

- Although language models can't draw, they can provide code for common diagram description languages (PlantUML, Mermaid, etc.).
- Language models can be linked to *agents*. For example, a calculation agent analyzes the text and forwards the figures it contains to a calculation program. The language model packs the result back into an answer in text form. Agents can also evaluate files attached to a prompt, try out code and search for errors contained therein, and so on. Agents are therefore an established tool for connecting a language model with external functions.
- Using RAG, as mentioned earlier, a language model can be linked to a constantly updated knowledge database. We describe this procedure in more detail in Chapter 12.

- Multimodal models move away from pure text processing and expand the communication basis to include sound and images. Special AI tools (*not* language models!) can even edit images and videos.

All of this takes us further and further away from the central topic of this book—the application of language models to code. As long as it's only about code, language models work amazingly well. Programming languages have a much clearer, more unambiguous syntax compared to "real" languages. This helps with the training of language models.

However, you should always be aware of the conceptual limitations of language models. Language models aren't really intelligent. Instead, they imitate human speech behavior on the basis of their training or generate code that is structured similarly to code from the training material.

Imitation Intelligence

The software developer Simon Willison, who works intensively with AI and language models in particular, uses the term *imitation intelligence* to deny language models genuine intelligence. (If you have three quarters of an hour to spare, you should listen to the Willison's presentation, "Imitation Intelligence" at *https://simonwillison.net/2024/Jul/14/pycon*.) However, just because a tool is flawed, that doesn't mean it's not useful. For comparison here, let's consider spellcheck. This often works well, but anyone who writes text from time to time knows the control function is far from perfect. The application of language models to coding is similar: if you understand the limitations and expect occasional errors, then there is still one tool left that you shouldn't do without.

1.5.9   The Size of the Context Window

The generation of code by AI tools in particular sounds scary and implausible. How can statistical procedures decide the content of the next line? The process works for several reasons:

- The language model was trained with millions of lines of code. It therefore knows common programming languages, common functions and algorithms, and so on.
- The syntax of programming languages is much simpler than that of "real" languages. In English, you can express one and the same thing in an almost infinite number of ways, linguistically more or less elegant, in short sentences or in convoluted constructions with foreign words, irony, cynicism, and so on. In this respect, coding for language models is the simple special case.
- Finally, language models not only take into account the actual question but also the connection with previous questions and answers as well as with the code already

available. All of this data forms the context for the continuation of the answer. Accordingly, the clearer you formulate your question, the more meaningful variable and function names your existing code uses, and so on, the better results you'll get.

If you're editing a file of a large software project, it would actually be ideal if the language model had the content of *all* code files available as context information. The language model would then know all classes with their methods and properties, the structure of the database schema, CSS attributes, and so on. Newly generated code would fit perfectly with the existing project.

Unfortunately, this isn't working at the moment. The size of the context memory (often called *context window* instead) is very limited. While the language model itself is many gigabytes in size, the context memory has room for just a few thousand tokens. The function of the context memory becomes clearer with an analogy to a computer: Even if you have a huge, terabyte-sized database on your computer's SSD, you're limited by the size of your computer's RAM when processing data.

Here's the obvious question, of course: Why doesn't the context memory simply get enlarged? In fact, this has already been done in recent months. GPT-4o supports a context memory of up to 32,000 tokens, which corresponds to around 60 pages of text. That is quite a lot. Other language models can handle even more context (200,000 tokens for Claude, 100 million tokens for Magic AI's experimental language models). But now, new problems are arising:

- The larger the context data, the longer it takes to respond to prompts. This effect is particularly dramatic in the local execution of LLMs.

- As the context size increases, the quality of the answer decreases instead of increasing. The LLM tends to be confused by all the additional information. It's not possible to separate essential from nonessential information (*increased noise*). The self-attention mechanism, which is an essential component of the GPT algorithms, also copes poorly with too much context.

- Another limit, the maximum length of the answer, comes to the fore. The *output token limit* is often much more restrictive than the context window limit and is only a few thousand tokens for many language models.

Of course, AI research is working on improving the context issues outlined here. But until that happens (if it happens at all), AI tools work best when the question is manageable and the amount of code to be processed is no longer than around 200 to 300 lines (as of early 2025). In this book, we've repeatedly encountered fundamental limitations of AI models, and they almost always had to do with a context window that was too small for the application (see also Chapter 10).

1.5.10   The Fill-in-the-Middle Approach

If you use language models via chat, the model simply responds to your questions (prompts). It can take into account the code passed to the prompt as well as the previous questions and answers of the session.

However, wizards such as GitHub Copilot must suggest usable code *without* a specific question. This can only work satisfactorily if the AI generator takes into account the code above *and* below the current cursor position as context.

This *fill-in-the-middle* (FIM) approach contradicts the purely forward-looking GPT model. It's not enough to find the most plausible continuation for a partially completed answer if the existing code underneath is ignored.

For FIM to work well, this use case must be taken into account as early as during the training of the language model. When creating the language model, then not only is the continuation of text trained but also the insertion of text between existing text or code segments.

LLMs such as GPT-4o from OpenAI were trained simultaneously for *both* use cases, that is, both for chat-based operation (purely forward-oriented) and for use as a code or text wizard (FIM). In the case of smaller, freely available models for local use, on the other hand, there are often different variants of the model that have been explicitly optimized for specific use cases. Accordingly, with GitHub Copilot alternatives such as Continue, you can specify *two* local models, one for chat tasks and a second for the code wizard that uses the FIM method.

Chapter 2

Pair Programming

Pair programming means that you create the code together with an artificial intelligence (AI) tool. It's rarely a good idea to prompt the AI tool with statements like, "My task is such and such, give me the solution!" This only works for simple problems or relatively short functions that are self-contained.

Most of the chapters in this book are intended for programmers or developers who are already reasonably proficient in their craft. We therefore assume that you're familiar with at least one programming language and know how to use functions, methods, and classes. Ideally, you also have some IT background knowledge, for example, you can work with databases or Git.

We make an exception to this rule in this chapter because here we specifically address those of you who are in the process of learning a programming language. In the examples, we demonstrate how AI tools can help you.

But although we focus on the learning process in this chapter, pair programming is, of course, also suitable for professionals! The greater your prior knowledge and the more complex your project, the more important it is that you proceed step-by-step; that is, specify the structure of the program and solve or improve one detail at a time with the help of AI. So, the aim of this chapter isn't to show you how an international bank account number (IBAN) verification function or a sudoku solver work. Rather, these sample applications are intended to give you examples of a reasonable use of AI tools. We want to show you what prompt engineering looks like in everyday work.

Our examples use the widely used Python and Java languages. But if you're learning another language, it doesn't matter! The prompts presented here work just as well in other languages. As in the other chapters of this book, it's not primarily about the answers provided by the AI tool (which we've often not presented at all or only in abbreviated form), but about how you can ask the AI tool questions in the most effective way.

The Ideal AI System

We've used ChatGPT with GPT-4o throughout for the examples in this chapter. However, you can just as easily use one of the competing chat systems. We had a particularly good experience using Claude from Anthropic while working on this book (*https://claude.ai/chat*). At the time of our tests in early 2025, Claude not only provided excellent quality answers but was also a little less verbose. The shorter answers are often easier to digest, especially if you only have a modest knowledge of programming. Even though ChatGPT is currently the best known, you can see there are exciting alternatives that—when not used extensively—are also free of charge.

If you're learning to program and don't want to complete a task as quickly as possible, then we strongly advise against using a coding wizard such as GitHub Copilot. Typical exercises such as "Write a function that tests whether the parameter is a prime number" or "Develop a class to map a bank account" are trivial for code wizards. All you have to do is enter a plausible function name such as testIfPrime, and the AI assistant will return the entire code. That means you'll finish quickly, but learn little.

Chat-based systems are better suited for learning. Ideally, you should explicitly state in the first prompt of a chat history that you're a beginner who needs some help and food for thought, but no finished code.

2.1 Structuring Code into Functions

Even if we're not assuming years of coding practice in this chapter, we're not starting from scratch either. There are other books for that. We assume that you already know what variables, branches, loops, and so on are and that you can handle these language constructs.

Experience shows that with this level of knowledge, the next hurdle is to structure longer code into functions in a meaningful way. That is the point at which we start here.

One starting point is to simply hand your AI chat partner the (spaghetti) code you've written and ask whether it sees a way of splitting the code across multiple functions.

> **Prompt:** I wrote this Python code to estimate the quality of a password. Can you help me break the code down into functions? Please explain why and how.
>
> [50 lines of code]

We can tell you this much: The code will almost always look neater afterward. ChatGPT also suggests meaningful function and parameter names and takes care of the documentation of the code in case you've neglected it. Try the preceding prompt or a similar formulation with your own code! (By the way, we go into more detail about improving existing code with AI help in Chapter 4.)

2.1.1 Defining Functions Yourself

At first, the procedure outlined earlier is instructive, but sooner or later, you should be able to divide your code into subfunctions yourself. As long as you find this difficult, you can ask for specific help.

Note the structure of the prompt: It begins with a relatively detailed summary of background information with regard to which programming language, which objectives, and which data structures you used. In addition, you should try to make it clear to your AI chat partner that it shouldn't spit out the finished code straight away (which would be easy for it), but that you want explicit support in structuring the code.

> **Prompt:** In my Python script, I have a nested list of numbers. I need to test if this list is a "magic square" where the row, column, and diagonal sums are all equal. Please help me *only* find a structure for the code. How should I define and use functions? Please also explain parameters. Don't provide the code; I want to write it myself.

To structure the code for checking if a nested list of numbers is a magic square, you can break down the problem into several smaller functions. Here's how you can define and use functions, along with an explanation of their parameters:

- Function to Check Square Shape:
  - Function Name: `is_square`
  - Parameters: `matrix` (a nested list of numbers)
  - Purpose: Ensure that the input is a square matrix (i.e., it has the same number of rows and columns).
- Function to Calculate Row Sums (. . .)
- Function to Calculate Column Sums (. . .)
- Function to Calculate Diagonal Sums (. . .)
- Function to Check Uniform Sums (. . .)
- Main Function to Check Magic Square (. . .)

[more details]

The answer is an excellent starting point for implementing the program step by step.

2.1.2   Defining Functions and Parameters

The starting point for the second example is an exercise. You're supposed to program a function that removes duplicates from a list. Your first problem isn't so much the code itself, but rather the definition of the function and the resulting data flow. Obviously, you'll pass the list to be processed as a parameter to the function. But how should the function return the result?

In the prompt, you again summarize the task and specify which programming language you're using. Again, what is important here is that you're not interested in the code, but in assistance with the declaration of the function and the structure of your code.

> **Prompt:** I need to write a Python function to remove duplicates from a list. How should I define the function? Don't provide actual code for the function.

To define a function for removing duplicates from a list in Python, you can follow these steps to structure the function:

- Use a clear and descriptive name, such as `remove_duplicates`.
- Parameters: Only `lst`, the list from which you want to remove duplicates.
- Return Value: The function should return a new list with duplicates removed.
- Method to Remove Duplicates: You can use different methods to remove duplicates, such as using a set or iterating through the list and checking for existing elements. Consider how you want to handle the order of elements (preserve or not).
- Edge Cases: Consider edge cases such as an empty list or a list with all identical elements.

[detailed explanation with more details]

Following the preceding, already shortened answer, ChatGPT can't resist the temptation to deliver (almost) finished code after all. Aside from that, however, the answer is a good basis for writing your own code. We particularly liked the fact that the AI tool points out that your function also has to cope with special cases (empty lists).

Perhaps you've already used other programming languages in the past. For some languages, you can change the data transferred in the parameter. This can increase the efficiency of your code, especially with large amounts of data. What does that look like in Python?

> **Prompt:** Instead of returning the result, is it possible or even recommended to modify the data passed as a parameter to a function?

Yes, it's possible to modify the data passed as a parameter to a function in Python, especially if the data type is mutable, such as a list. However, whether it's recommended depends on the context and the specific use case. Here's a detailed explanation.

[two pages of details]

The answer is symptomatic: ChatGPT readily generates endless amounts of text to explore the advantages and disadvantages of different variants. In most cases, however, the AI tool avoids making a clear recommendation. In this case, almost everything speaks against a direct modification of the data, especially the readability of the code. Instead, it's better to follow the simple approach of processing parameters as an input and using a return to provide the result (output).

Depending on which language model you use, the answer to the same question may be completely different. The following lines were generated by the free Llama 3 language model, which we executed locally with Ollama. We liked this answer better, both in terms of content and commitment.

> In Python, it's generally not recommended to modify the original data structure.
>
> [more details]
>
> So, while it's technically possible to modify the input list, I would recommend against it.

We also had a go with the AI chat tool Claude from Anthropic. Its answer was similarly noncommittal as the one from ChatGPT, but contained a new aspect: Claude points out that it's sometimes useful to develop *two* versions of a function, one to process the data in place and the second to return results.

> [pros and cons of both approaches]
>
> Alternative approach: A common Python convention is to provide both versions:
>
> - An in-place function (e.g., `list.sort()`)
> - A function that returns a new object (e.g., `sorted(list)`)

However, the object-oriented approach, that is, the application of a method to a list, is aimed at fairly advanced Python fans.

2.1.3   Detailed Questions about Functions and Parameters

When dealing with functions and parameters, there are a number of special cases that vary greatly depending on the programming language. Of course, you can read the online documentation or a Python book, but you can also simply ask an AI tool. If you don't feel like reading lengthy digressions, you should ask for a concise formulation of the answer. ChatGPT in particular has a tendency to be verbose in its responses and to provide much more detail than you need.

Don't forget to include the name of the programming language you're using in the prompt. (You don't have to do this for follow-up questions, in which case, the language model already knows the correct context.)

The first question refers to the frequently occurring problem that a function should return a multipart result. Note the prompt again: to make it clear to the AI tool what we're talking about, we've also provided a simple example.

> **Prompt:** Can my Python functions return more than one data item (for example, a string and a number)?

> Yes, your Python functions can return more than one data item, such as a string and a number, using several methods. Here are the most common ways:
>
> ```
> # use a comma-separated list of values to create a tuple
> def get_info():
>     return "Hello", 42
> # return a list containing the values
> def get_info():
>     return ["Hello", 42]
> ```
>
> [three more variants: dictionary, namedtuple, custom object]

There are just as many questions for programming beginners about declaring the parameter list of functions. You can ask the AI tool of your choice about this like a fellow student or a tutor.

> **Prompt:** Using Python, how can I define a function to pass a variable number of parameters? And how can I process those parameters inside the function?
>
> **Prompt:** I want to use optional parameters with default values. What's the Python way to do this?
>
> **Prompt:** How should I name functions and parameters?
>
> **Prompt:** Within a Python function, can I access/change variables defined outside?
>
> **Prompt:** What is the recommended order for a function with several parameters?

ChatGPT is at its best when it comes to such questions and describes every conceivable variant and special case for pages on end. If you don't have the time or inclination to read through this, you need to make it clear to the AI tool that you're only interested in the basics and that you're still at the beginning of programming.

> **Prompt:** Once again, please. Be concise, drop special cases. I am a Python beginner.

You'll get the best answers if you not only ask detailed questions but also provide a specific use case or example.

> **Prompt:** OK. Now, consider only required parameters. What should come first? Suppose I want to write a function to find and replace items in a list. What's better, replace(data, find, replace) or replace(find, replace, data)?

For functions with only required parameters, the general guideline is to order parameters from the most important or the most specific to the least important or the most general [more details]. For a function that replaces items in a list, it's clearer and more intuitive to place the data last:

```
def replace(find, replace, data):
    # Code to replace items
    pass
```

Why this order? [more details]

2.1.4   Code Optimization

You may already have working code that you just want to optimize. ChatGPT can also help you with this. However, be sure to check after each step whether your program is still working as intended!

The first two prompts are about adapting existing code to the usual Python rules. ChatGPT already knows the conventions such as correct indentation and the naming of variables and functions. You only need to go into these details if you have special requests. Ideally, you should then provide templates and specific statements.

Prompt: Please adapt this code to the Python naming conventions.

[100 lines of code]

Prompt: Could you find better names for the variables and functions used in this code?

Good examples are: `total_sales`, `current_temperature`, `user_input_calculcated`.

Don't use: `TotalSales`, `CurrentTemperature`, `UserInputCalculated`.

Many programmers are quite careless when it comes to commenting code. AI tools can also take over this unpopular work (see also Chapter 6). In the simplest case, a mini prompt as in the first example is sufficient.

Prompt: Please add comments to this code.

Depending on who is to read your code and comments, you should specify this target group. Note, however, that although the AI tool can usually understand your code, it can't guess *why* you've decided on a particular course of action. AI comments can therefore only partially replace your own comments.

Prompt: Add comments to this file to make the code understandable to a novice programmer.

> **Tip**
>
> Commenting code afterwards can also help if you've received code from a colleague, from the internet, or from another source and don't understand it properly.

Perhaps the function you've developed works but is very slow or takes up a lot of memory. If you ask an AI tool to optimize your code, you should be as specific as possible about what needs to be optimized (e.g., speed).

> **Prompt:** This function is rather slow. Do you see any way to make the code more efficient?

> **Prompt:** When processing large files, this function uses far too much memory. Is there a way to reduce the memory consumption of the code?

2.1.5   Code Reuse

AI tools shouldn't be your only aid when learning a programming language. If you're searching for code on the internet and come across constructions that you don't understand, you can have them explained to you. (Just as you should never blindly accept code generated by the AI unless you really understand it, this rule also applies to code from Stack Overflow or other reputable sites. The code may actually be correct but is intended for a completely different application than the one you have in mind.)

The following prompt shows how you can combine the strengths of the internet as a universal advisor with those of AI. Here, the AI tool is supposed to explain the linalg.norm function, which was previously unknown to you.

> **Prompt:** This function apparently calculates the distance between some points. Can you explain what np.linalg.norm() does?
>
> ```
> def get_distances(points_xy):
>     l1 = np.linalg.norm(
>         points_xy[0, :] - points_xy[1, :])
>     l2 = np.linalg.norm(
>         points_xy[1, :] - points_xy[2, :])
>     return l1, l2
> ```

You may be switching from one language to another, or you may have found a solution to your problem on the internet in another language. Most AI tools are able to translate your code from one language to another.

Prompt: Last year I wrote a PowerShell script to download the image of a web page. How can I rewrite this code using Python?

[20 lines of PowerShell code]

In our experience, such code conversions don't always work perfectly. Every programming language, every library has its own peculiarities. Take the time to test the code thoroughly and scrutinize its functions. If you want to learn programming and the focus isn't on completing a task quickly, it's even better to use the modules or libraries suggested by the AI bot as a starting point for your own research and then develop the code yourself.

2.1.6   Splitting Code across Multiple Files

For smaller programs, splitting the code into independent functions is sufficient to ensure that the code remains readable, clear, and maintainable. With complex applications, however, there comes a point where it's advisable to distribute the code across multiple files.

In the following prompt, we asked the AI tool to break down a 250-line code file into subfunctions. We weren't only interested in the completion of the work itself, we also wanted to know why the AI tool chooses a certain approach.

Prompt: I have written a long Python script, and I want to split the code in a logical way across more than one code file. Can you help? Please explain why and how.

[250 lines of code]

To split your Python script logically across multiple files, we'll use a modular approach. The goal is to organize your code into coherent units that make it easier to maintain, understand, and test. Here's a step-by-step breakdown of how to do it:

- Identify Logical Units [more details]
- Create Modules [more details]
  - This is the new code structure:

```
project_directory/
|-- __init__.py
|-- config.py
|-- logger.py
|-- auth.py
|-- web_utils.py
|-- main.py
```

- Split the Code [more details]

- Import the Modules: In `main.py`, you import the necessary functions from each module to keep the script organized and modular.

Benefits of modularization [more details]

In the sample code we used, ChatGPT has actually found a sensible way of distributing the code across multiple files, each of which can be easily extended. But perhaps you're not yet familiar with the concept of Python modules. You've probably already used ready-made modules, but haven't yet defined your own.

> **Prompt:** Looks good. However, I am not yet familiar with creating modules. I have used Python modules before, but I have never written modules myself. Can you explain the basic concepts of how to create my own modules? Be concise!

The answer leaves nothing to be desired. However, ChatGPT immediately goes into packages as well. If you don't know what this is (in principle: multiple modules in one directory), you should ask a follow-up question.

> **Prompt:** What's the difference between a module and a package?

If you now take the time to try out the module and package examples suggested by ChatGPT yourself and perhaps expand them here and there, you'll be familiar with the concept an hour or two later. Great!

Tips

There are major differences between common programming languages when it comes to organizing code across multiple files, so—one more time—be sure to specify the language in which you're programming!

Also note that many AI tools have problems processing very long code. While splitting 250 lines of code across over three or four files works wonderfully, ChatGPT and others often fail if the code size exceeds 500 or 1,000 lines. So, don't wait too long before disassembling your code! The reason for the failure with long code files are limits for the size of the context window and for the length of the response (see also Table 1.1 in Section 1.2).

2.2   Example: IBAN Validation

For this example, you're supposed to develop a Python function to check the plausibility of IBANs. IBANs must comply with three rules:

- The first two digits contain the country code (e.g., FR for France).
- The total number of characters varies depending on the country (e.g., 22 characters in Germany).
- In addition to the actual account number, the IBAN contains two check digits. A checksum can be calculated using a formula.

You can read the details in Wikipedia. Of course, there are also plenty of ready-made code samples for IBAN validation on the internet. But the purpose of the exercise is to develop suitable code yourself.

In this section, we've summarized prompts that can help you along the way. We've refrained from printing the answers and the resulting code in all stages of program development. Our aim is to give you tips on how to use prompts to arrive at a solution step-by-step, with as much code as possible coming from you. (If you're interested in the code, you'll find it in the sample files for this book.)

2.2.1 Start

The start prompt not only outlines the problem but also describes your specific situation. What you want is learning support, not code for a ready-made solution. On the contrary, you want to program as much code as possible yourself.

> **Prompt:** I am a Python beginner and want to learn programming. I need to write a function to check if a string contains a valid IBAN code. Please help me, but don't give me a complete solution. Instead, give me hints on how to get started (and how to proceed later when I ask for more details).

As intended, the ChatGPT-4o language model used in our tests doesn't provide the finished code, but it does provide a lengthy response. However, all we wanted was a starting point! Let's make a new attempt.

> **Prompt:** Excellent, but this is too much at once. Please try again, and make your answers no longer than one page (about 2,000 characters).

> Of course! Let's take it step-by-step and focus on one part at a time. Here's how you can start. Each country has a specific IBAN length. Your first step is to check if the IBAN has the correct length for its country. First, create a dictionary with country codes and their corresponding IBAN lengths.
>
> ```
> iban_lengths = {
>     'DE': 22,  # Germany
>     'GB': 22,  # United Kingdom
>     # Add more countries as needed
> }
> ```

Then write a function to check the length. Once you confirm that the IBAN length is valid for the given country, you can proceed to the next step. When you're ready to continue or have any questions about this part, feel free to ask!

2.2.2   Checking the IBAN Length

The AI tool therefore suggests checking the length of the IBAN code first. For this exercise, it isn't necessary for our code to work for *all* countries. We therefore limit ourselves to three countries and ask the AI tool for the prescribed length. ChatGPT packs the answer directly into a Python dictionary.

Prompt: Can you provide IBAN lengths for AT, DE, and CH?

```
iban_lengths = {
    'AT': 20,   # Austria
    'DE': 22,   # Germany
    'CH': 21,   # Switzerland
}
```

We adopt the dictionary proposed by ChatGPT. The task now is to extract the first two characters from the string and extract the desired length from the dictionary. We have two detailed questions on this:

Prompt: How do I extract the first two characters of a string?

Prompt: How can I test if the `iban_length` dictionary contains an entry for my country?

2.2.3   Troubleshooting

This information is sufficient for us to write a first basic version of the program, which simply checks the length of the IBANs. However, Python suddenly reports an error in our code that we don't understand. We ask ChatGPT for advice.

Prompt: My editor says this code, which is part of a function, is wrong. Why?

```
if not country_code in iban_lengths:
    print ('Country code not found')
    return False
else if len(iban) != iban_lengths[country_code]:
    print ('IBAN length is incorrect')
    return False
else:
    return True
```

The AI tool immediately recognizes the error (instead of else if, it should read elif), but it's annoying with a one-and-a-half-page answer with lots of irrelevant details.

We're continuing to tinker with our code and have the impression that it works—with the exception of the checksums that haven't yet been implemented. At this point, the code looks as follows:

```python
#!/usr/bin/env python3

iban_lengths = {
    'AT': 20,  # Austria
    'DE': 22,  # Germany
    'CH': 21,  # Switzerland
}

# Check the length of the IBAN
def check_iban_length(iban):
    country_code = iban[:2]
    if country_code not in iban_lengths:
        print('Country code not found')
        return False
    elif len(iban) != iban_lengths[country_code]:
        print('IBAN length is incorrect')
        return False
    else:
        return True

# Verify the IBAN; add checksum verification later
def validate_iban(iban):
    return check_iban_length(iban)

# Main code starts here
iban = 'AT4832000000123458642'
if validate_iban(iban):
    print('Valid IBAN')
```

2.2.4   Test

Developing your own program also involves verifying the correct behavior of the program using a range of correct and incorrect test data. To do this, you want to pass some IBANs to your function in a loop. If everything works, your function evaluates all correct IBANs as correct and all incorrect IBANs as incorrect. To do this, you need test data, which ChatGPT can provide you with. (Chapter 5 deals intensively with the question of how you can use AI tools to create *unit tests*, that is, test functions that verify the correct functioning of a program.)

Prompt: I want to test my function with several IBANs. Can you provide a list with three valid and three invalid codes for AT, DE, and CH?

You can test your function using the provided IBAN codes. Here's a sample testing script:

```
def test_iban_validation():
    valid_ibans   = [ 'AT611904300234573201', ... ]
    invalid_ibans = [ 'AT611904300234573201', ... ]

    for iban in valid_ibans:
        assert validate_iban(iban), \
                f"Valid IBAN failed: {iban}"

    for iban in invalid_ibans:
        assert not validate_iban(iban), \
                f"Invalid IBAN passed: {iban}"

    print("All tests passed!")
```

ChatGPT not only provides two lists of valid and invalid IBANs but also a test function that passes these codes to our IBAN validator one after the other. ChatGPT's excessive zeal isn't in our interest, but where the code is already there, we want to use it.

If an unexpected result occurs in both loops, an error is triggered via assert. However, we've never heard of assert, so we ask ChatGPT for an explanation.

Prompt: What is assert good for?

2.2.5   Removing Spaces from the IBAN

A little smarter than before, we next add a function to our code that removes spaces from the IBAN and converts all lowercase letters to uppercase. However, calling up our functions seems cumbersome to us. That's why we ask our AI learning partner if it has a better suggestion for combining the necessary function calls.

Prompt: Is there a more elegant way to write this function?

```
def validate_iban(iban):
    # remove spaces
    iban = tidy_iban(iban)
    # test length
    result = check_iban_length(iban)
```

```
# return result of check_iban_length()
return result
```

Yes, there is a more elegant way to write the `verify_iban` function by streamlining the code and making it more Pythonic [one page of unnecessary details].

```
def validate_iban(iban):
    return check_iban_length(tidy_iban(iban))
```

The nested function call reduces three lines to just one statement, which is still easy to read.

2.2.6   Checksum Test

Now let's tackle the most difficult piece of the puzzle, the checksum test. We've read through the relevant instructions in Wikipedia, but we find it difficult to implement them in code. We therefore try to proceed in small steps. To slow down ChatGPT's flow of speech a little, we preemptively add "be concise" to the prompt.

> **Prompt:** My code now cleans up the IBAN string and checks for correct length. Next, I am trying to do a checksum test. According to Wikipedia, I should move the country code and the next two digits to the end of the string. Can you help me with this step? Please try to be concise.

Unfortunately, ChatGPT again provides an answer that extends over two pages. We first have to look for the line that is decisive for us:

```
rearranged_iban = iban[4:] + iban[:4]
```

Now we need to replace the country code letters in this string with two digits each, A with 10, B with 11, and so on. For a change, we use Stack Overflow to help us. But the one-line code presented there is difficult to digest. We ask ChatGPT to explain the statement to us.

> **Prompt:** I found this code to replace letters by numbers, but I don't understand it. Can you explain it?
>
> ```
> iban = ''.join(str(ord(c) - 55) if c.isalpha() else c
>             for c in iban)
> ```

The explanation makes things a little clearer, but not entirely. We therefore decide to formulate the code ourselves in an "ordinary" `for` loop, perhaps not quite typical of

Python, but easier to understand. To do this, we ask ChatGPT for help again, asking very specific detailed questions so that we can assemble the code ourselves in tiny steps.

> **Prompt:** How can I loop over every character of a string? (Be concise!)
>
> **Prompt:** How can I test if a character is a digit?
>
> **Prompt:** How can I replace the country code characters with the required numbers for the checksum test (i.e., replace A with 10, B with 11, etc.)?
>
> **Prompt:** Now I have a long string of numbers. How do I convert it to an integer?
>
> **Prompt:** Is there no risk of an overflow error? The numbers are quite large.

After an hour's work, our code looks as follows:

```
# test the checksum
def test_checksum(iban):
    iban = iban[4:] + iban[:4]
    digit_string = ''
    # loop over the IBAN and replace letters with numbers
    for c in iban:
        if c.isdigit():
            digit_string += c
        else:
            digit_string += str(ord(c) - 55)
    # convert the string to an integer and test if modulo 97 is 1
    return int(digit_string) % 97 == 1
```

2.2.7   Trouble with the Test Data

We actually have the impression that our code works. However, an error occurs with one of the test IBANs supplied by ChatGPT. Of course, we first look for the error in our code, but ultimately, we become suspicious. We ask our AI partner whether one of the test IBANs provided to it might be incorrect.

> **Prompt:** My code seems to be working now. However, one of the "valid" IBANs you provided for testing is considered invalid by my program. Is CH3109000000532432318 really a valid IBAN?

ChatGPT must admit that this account number is indeed incorrect. That's pretty annoying as we had already spent some time troubleshooting!

2.2.8   Code Optimization

Finally, we ask ChatGPT whether the tool still sees room for improvement in the finished code (see the `iban3-final.py` file in the sample files for this book).

> **Prompt:** This is my final code. It works well. Do you see any room for improvement?
>
> [60 lines of code]

Of course, ChatGPT has a few suggestions for improvement (see `iban4-optimized.py`). Among other things, it converts our loop for processing the country code back into the hard-to-read but more efficient `''.join(...)` structure that we had already found on Stack Overflow, adds a `try-except` construction for error protection, and builds in clearer error messages.

Overall, our experiment was satisfactory. Unfortunately, ChatGPT consistently ignored the repeated request for compact, short answers during our tests. The actual answer was followed by additions, summaries, or more information that we didn't want to receive at this point. Other than that, the AI tool was a great help both in the actual programming of the function and in our desire to develop and understand the code step by step.

2.3   Object-Oriented Programming

Structuring code into functions or methods is only the first step on the way to the well-thought-out development of complex, large programs. It's even more difficult to design code in a meaningful object-oriented way. What belongs to a class? Which data is public, and which is private? What details does the constructor have to take care of? At which (data) boundaries are classes differentiated from one another? Where can mechanisms such as inheritance or interfaces be used appropriately?

As in the previous sections, our aim here isn't to teach you how to program. Nor did we consider it appropriate to print a lot of code here. Our primary goal is to use examples to motivate you to let AI bots or wizards provide you with targeted support. Not only does this work with regard to formulating the code, but also—and even more important—in the correct designing of classes.

> **Why Java?**
> The following examples use Java as the programming language. Although you can also define classes in Python, its strengths lie elsewhere. Java offers significantly more object-oriented features.

Admittedly, Java is a fairly old language. You can formulate object-oriented code much more elegantly in new programming languages such as Swift or Kotlin. However, these languages are relatively new, and features change or are added. For this reason, and because of the limited distribution (i.e., the limited availability of training material), language models work better for long-established languages than they do for new languages.

Long story short—even if our love for Java is limited, this widely used language seemed to be the best choice for the following examples.

2.3.1 Example: User Account Class

The following prompts are intended to show how AI tools can help you with the object-oriented design of programs. The first prompts concern a program section in which a user account is represented by a Java class. In the prompt, we summarize the objective of the class, list its most important properties, and name the programming language we're using. Only then does the actual question follow: What kind of validation should be performed in the constructor of the class?

Prompt: I need to write a Java class for a user account. Important attributes are account name, full name, contact email, password hash, and a list of recent login times. What kind of validation should I do in the constructor?

When creating a Java class for a user account with attributes like account name, full name, contact email, password hash, and a list of recent login times, it's important to include appropriate validations to ensure data integrity and security. Below is a suggested Java class with the required validations:

```java
import ...;

public class UserAccount {
    private String accountName;
    private String fullName;
    private String contactEmail;
    private String passwordHash;
    private List<String> recentLoginTimes;
    // Constructor
    public UserAccount(String accountName, ...) {
        this.setAccountName(accountName);
        this.setFullName(fullName);
        ...
    }
}
```

```
    // Validation methods used by constructor
    private void setAccountName(String accountName) {
        if (accountName == null ||
            accountName.isEmpty() ||
            accountName.length() < 3 ||
            accountName.length() > 20)
        {
            throw new IllegalArgumentException(
              "Name must be between 3 and 20 characters.");
        }
        this.accountName = accountName;
    }
    private void setFullName(String fullName) { ... }
    private void setContactEmail(String contactEmail) { ... }
    private void setPasswordHash(String passwordHash) { ... }
    private void setRecentLoginTimes(
                    List<String> recentLoginTimes) { ... }

    // Utility method to validate email format
    private boolean isValidEmail(String email) { ... }

    // Getters
    public String getAccountName() {
        return accountName;
    }
    public String getFullName() { ... }
    public String getContactEmail() { ... }
    public String getPasswordHash() { ... }
    public List<String> getRecentLoginTimes() { ... }
}
```

Validation Details

- **Account Name:** Must be between 3 and 20 characters, shouldn't be null or empty.
- **Full Name:** Shouldn't be empty and must be less than 50 characters.
- **Contact Email:** Must be a valid email format using a regular expression.
- **Password Hash:** Must be a 64-character string assuming a SHA-256 hash, shouldn't be null or empty.
- **Recent Login Times:** Initializes an empty list if the provided list is null to avoid Null-PointerException.

[more details]

As is so often the case, the answer goes a little beyond the actual question. ChatGPT proposes the complete structure of the class, defines the private instance variables, and provides getter and setter methods for data access. The validation takes place in the setter methods. We're questioning this concept.

> **Prompt:** Java: Using getters and setters only seems to complicate my code. What's the benefit?

In its response, ChatGPT explains that the chosen approach actually makes sense here. We therefore follow the suggestion and implement the class, including properties, getter and setter methods, and the constructor.

2.3.2   Output of the Object

We next want to equip the class with the toString method, which is particularly useful for tests because it allows us to easily output an object with all of its properties.

In the prompt, we ask ChatGPT not to output the code of the entire class each time, but only the code of the new toString method. We've greatly shortened the code outputs. If you want to view the entire code, just take a look at the sample files for this chapter.

> **Prompt:** Please add a toString method to easily output the account data in a readable form. Don't answer with the entire code; only show the necessary changes.

Here are the necessary changes to add a toString method for the UserAccount class:

```
@Override
public String toString() {
    return "UserAccount{" +
            "accountName='" + accountName + '\'' +
            ...
            ", recentLoginTimes=" + recentLoginTimes +
            "}";
}
```

2.3.3   JSON Export and Constructor

Now we want to extend the class with a method for the JavaScript Object Notation (JSON) export. We also want to implement an additional JSON constructor. In the prompt, we ask for a solution that is as simple as possible without the use of external libraries.

> **Prompt:** I want to export all attributes as a JSON string. Can you provide a simple method without using an external library?

To export the `UserAccount` object to a JSON string without using any external libraries, you can manually construct the JSON string. Here's how you can do it:

```
// Method to convert the object to JSON string
public String toJson() {
    StringBuilder json = new StringBuilder();
    json.append("{");
    json.append("\"accountName\":\"").
        append(escapeJson(accountName)).
        append("\",");
    ...
    return json.toString();
}
// Helper methods
private String escapeJson(String value) { ... }
private String listToJson(List<String> list) { ... }
```

Once we've tested the code, we want a JSON import function as a second constructor in the next prompt.

Prompt: What might a second constructor that expects a JSON string look like?

To create a second constructor that initializes a `UserAccount` object from a JSON string, you need to parse the JSON string and extract the necessary fields. Because we're not using any external libraries, we'll manually parse the JSON string. Here is how you can implement the second constructor:

```
// Second constructor expecting a JSON string
public UserAccount(String jsonString) {
    this(parseJson(jsonString, "accountName"),
        ...
        parseJsonArray(jsonString, "recentLoginTimes"));
}
// Helper methods to parse JSON strings and arrays
private static String parseJson(
  String jsonString, String key) { ... }
private static List<String> parseJsonArray(
  String jsonString, String key) { ... }
```

2.3.4   Example: Bank Account

The next exercise is about mapping a bank account as a Java class. In the prompt, we first describe the task. But before we get started, we still have some basic questions.

Which data types should we use? Should the properties be implemented as private or public variables?

> **Prompt:** I have to write a Java class for a bank account. It stores the name of the account holder, the current balance, and a list of all previous transactions. Which data types should I use? Should I make my instance variables public or private?

To write a Java class for a bank account, you need to decide on appropriate data types for each attribute and determine the visibility of your instance variables. Here are the recommendations for both [more details].

Instance variables should generally be private to encapsulate the data and protect it from unauthorized access or modification. Public access should be provided through getter and setter methods.

```java
import java.math.BigDecimal;
import java.util.ArrayList;
import java.util.List;

public class BankAccount {
    private String accountHolderName;
    private BigDecimal currentBalance;
    private List<Transaction> transactionHistory;

    // constructor
    public BankAccount(String accountHolderName) { ... }

    // methods
    public void deposit(BigDecimal amount) { ... }
    public void withdraw(BigDecimal amount) { ... }
    public String getAccountHolderName() { ... }
    public BigDecimal getCurrentBalance() { ... }
    public List<Transaction> getTransactionHistory() { ... }

    // nested class representing a transaction
    public static class Transaction {
        private BigDecimal amount;
        private String type;
        private String date;
        ...
    }
}
```

ChatGPT opts for private class variables, which is obvious in view of the question. During our tests, ChatGPT was also overzealous in providing methods for securely debiting and crediting money. In particular, the `withdraw` method refuses to debit more money than is currently available. The `deposit` method checks that the amount of the deposit is greater than 0.

We wanted to ask about these details in the next step, but unfortunately, this is bypassed. If you want to program efficiently and quickly, then ChatGPT's comprehensive answers are undoubtedly useful. If, on the other hand, you want to learn to program and contribute as much of your own code as possible, the overzealousness of AI tools is a hindrance.

2.3.5   Example: Sorting Photos Based on EXIF Data

The third example is formulated even more openly. The goal is a program to sort photos into directories according to the dates they were taken. The prompt describes in great detail what the program should be able to do and which library should be used to read the exchangeable image file format (EXIF) data. Once again, we try to make it explicitly clear to our AI tool that we don't want finished code, but a suggestion for organizing the program.

> **Prompt:** I am trying to write a Java program that loops through all the JPEG files in a directory. The program uses `metadata-extractor` to read the EXIF data of each file. Images with valid EXIF data are then moved to a new directory `yyyy-mm`, where `yyyy` is the year of creation, `mm` is the month.
>
> Please help me organize the code in a proper object-oriented way. Just provide a framework for the code, no methods yet.

Certainly! To organize your Java program in a proper object-oriented way, we can create a few classes to handle different responsibilities. Here's a framework with a brief explanation of each class's role:

- `JPEGProcessor`: Main class to initiate the process and handle the overall flow.
- `FileHandler`: Responsible for file operations like reading and moving files.
- `EXIFExtractor`: Uses `metadata-extractor` to read and parse EXIF data from JPEG files.
- `DirectoryOrganizer`: Manages the creation of new directories based on EXIF data.
- `Logger`: Handles logging of operations and errors.

[more details with code for all classes]

Of course, there are many ways to distribute the necessary code across several classes. The AI proposal summarized here is absolutely useful and a good starting point. What's annoying is that ChatGPT again can't resist the temptation to include a lot of sample code.

2.3.6   Example: Dealing with Singletons

When you learn a new programming language—regardless of whether it's your first language or whether you can already program in other languages—you'll always come across new features. You'll see programming techniques in books and on the internet that you don't understand straight away.

Pair programming with an LLM allows you to try out concepts in dialog mode and ask specific questions based on your task. The more information you provide about your level of knowledge and the initial situation, the better the AI can respond to your question.

In the following prompt, we assume that you've heard or read about the singleton pattern. Although the concept sounds useful, what does its implementation in Java look like?

Prompt: How can I realize the singleton pattern in Java?

ChatGPT interprets the question in the sense of determining how many possibilities there are to implement singletons in Java, and it presents six variants in its answer. We've only reproduced the first one.

In Java, the Singleton pattern ensures that a class has only one instance and provides a global point of access to it. There are several ways to implement a Singleton pattern in Java, each with its own advantages and disadvantages. Here's a detailed look at the most common implementations.

Eager Initialization: This approach creates the instance when the class is loaded. It's simple but may lead to resource wastage if the instance is never used.

```
public class Singleton {
    // private static instance, so it can be accessed
    // by only getInstance() method
    private static final Singleton instance =
      new Singleton();
    // private constructor to prevent instantiation from
    // other classes
    private Singleton() {}

    // public method to provide access to the instance
    public static Singleton getInstance() {
        return instance;
    }
}
```

[more details: lazy initialization, thread-safe singleton, thread-safe singleton with double-checked locking, Bill Pugh singleton implementation, enum singleton]

ChatGPT overshoots the mark—as usual—and bombards you with details of six(!) implementation variants. That goes too far because you intend to use singletons to store a database connection centrally in the program. The concrete prompt leads to a much more concrete answer.

Prompt: Which variant is best if my program needs a single database connection?

If your program needs a single database connection, the Bill Pugh Singleton implementation is generally the best variant to use. [more details]

You trust the answer and refrain from comparing it with other implementation variants. Instead, in the next prompt, you ask whether it wouldn't be sufficient to equip your intended database class with a conventional constructor only.

Prompt: I intend to write a database class that contains all database functions I need for my program. What are the advantages of the Bill Pugh Singleton implementation over a traditional constructor?

Once you're finally convinced, you should further specify the objective in the next prompt. You specify which database system you want to use and ask the AI tool for a simple sample implementation.

Prompt: OK, let's use the Bill Pugh Singleton implementation. My database system is PostgreSQL. Show me a simple implementation. Be concise!

2.4 Example: Quiz

In this more extensive example, we assume that you want to program a simple guessing game. The program selects a few questions at random from a pool of questions. It asks the questions and provides several multiple-choice answers marked with the letters a, b, c, and so on to choose from. The player enters a character string (e.g., ad if the first and fourth answers are correct), and the program analyzes the answer and asks the next question.

2.4.1 Class Design

This section is primarily concerned with organizing the classes required for the program. You're still inexperienced when it comes to class design, so you ask an AI tool for help. In the first, very detailed prompt, you outline the basic functions of the program and name the programming language.

You immediately make it clear to the AI tool that there are multiple-choice questions where *multiple* correct answers are permitted. This is important information for the design of the program. Without this information, ChatGPT will most likely assume that there is only one correct answer per question. If you develop the program under this assumption at this point, extensive changes will be necessary later to extend the functionality. It's much better to clarify this detail now.

> **Prompt:** I would like to write a small quiz game in Java. There is a predefined pool of questions. When the program starts, the player has to answer five questions randomly chosen from the pool.
>
> For each question, there is a set of possible answers (min. 2, max. 6). At least one answer has to be correct. Some questions may have more than one correct answer. In this case, ALL correct answers must be selected.
>
> Please help me design the necessary classes. Don't provide any finished code yet. Just help me decide on good class names and define the right instance variables. Leave constructors and methods for later.

Designing a quiz game in Java requires clear and well-structured classes to manage questions, answers, and the gameplay itself. Below is a proposed class design with suggested instance variables:

- **QuizGame:** Manages the overall game flow and logic. Instance variables:
  - `List<Question> questionPool`: Holds all possible questions for the game.
  - `List<Question> selectedQuestions`: Stores the randomly chosen questions for the current game.
  - `int score`: Tracks the player's score.
  - `int currentQuestionIndex`: Keeps track of the current question in the game.
- **Question:** Represents a single quiz question. Instance variables:
  - `String text`: The text of the question.
  - `List<Answer> answers`: A list of possible answers for this question.
  - `List<Answer> correctAnswers`: A list of correct answers.
  - `int id`: Unique identifier for the question (optional, for easier management).
- **Answer:** Represents a possible answer to a quiz question. Instance variables:
  - `String text`: The text of the answer.
  - `boolean isCorrect`: Indicates whether the answer is correct.
  - `int id`: Unique identifier for the answer (optional, for easier management).
- **Player:** Represents the player in the game [. . .].
- **QuizUtil:** Utility class for helper methods (e.g., random selection, validation . . .)

[detailed description]

The suggestion is definitely useful. You decide to initially organize the code into three classes: Main, Question, and Answer. You start the implementation independently, set up the start code in Main, and program the trivial Answer class.

2.4.2 Implementing the Question Class

In the next prompt, you request support with programming the Question class. It's important that you tell your AI tool what code already exists because it has to reference it in the Question class.

> **Prompt:** OK. I created a Main class to run/test the program. I also wrote the Answer class like this:

```
public class Answer {
    public String text;
    boolean isCorrect;

    // constructor
    public Answer(String text, boolean isCorrect) {
        this.text = text;
        this.isCorrect = isCorrect;
    }
}
```

> Next, I want to implement Question. Please suggest initial code (only instance variables and the constructor, without the optional id variable).

```
import ...;

public class Question {
    public String text;
    public List<Answer> answers;
    public List<Answer> correctAnswers;

    // Constructor
    public Question(String text,
                    List<Answer> answers,
                    List<Answer> correctAnswers) {
        this.text = text;
        this.answers = answers;
        this.correctAnswers = correctAnswers;
    }
}
```

2.4.3 A Pool of Questions for Testing

Before you look at the other methods of these classes, you need test material, that is, some quiz questions. Creating them is a rather dreary task that can be delegated to ChatGPT and others.

> **Prompt:** As a base for testing the code, please provide a method for the `Main` class to initialize a pool of 10 geographic questions.

ChatGPT initially only provides questions with exactly *one* correct answer. The AI tool only adds questions with multiple matching answers on request.

> **Prompt:** Perfect. However, you could include a few questions with more than one correct answer (e.g., rivers in Brazil)?

2.4.4 Methods for the Question Class

Now that we have the first quiz questions, it's now time to design the `Question` class. At least two methods are required here: one for displaying a question together with its solution and for entering the answer, and the second for evaluating the answer. In the following prompt, you ask for the implementation of the first method, specifying relatively precisely how the question should be presented (the answers should be marked with lowercase letters) and in which format the program expects the answer. To prevent the AI tool from outputting the entire class code, we ask that it only returns the new method.

> **Prompt:** Add a method to the `Question` class to print the question text and all of the enumerated answers such as a), b), c), and so on. The method prompts the user for an answer string, for example, "ad" if answers a) and d) are correct. Provide code only for the method, not for the entire class.

The AI tool suggests two new methods: `askQuestion` implements the prompt's specifications and returns the answer as a character string. The unsolicited `validateAnswer` method evaluates this character string and checks whether it contains all applicable answers, but no further letters for nonapplicable answers.

In this example, the `validateAnswer` method is the trickiest. But before you could even think about how to program the evaluation of the character string yourself, ChatGPT has already presented a ready-made method. This is annoying.

The code created up to this point can be found in the sample files for this book. At this point, we leave this example. Of course, various methods are still missing to complete the program (game sequence, random selection of five questions, presentation of the questions, display of the result [e.g., three out of five questions answered correctly], etc.). Maybe you want to complete the example yourself—that would be a good exercise in the use of AI tools. Or, can we motivate you with this example to implement this or a similar game yourself with the help of AI? You decide to what extent you want to be supported.

Overzealous Learning Aid

Unfortunately, ChatGPT and others tend to be overzealous when it comes to their use as a learning aid. The AI tools often provide more code than necessary (even if you explicitly request the opposite in the prompt). As soon as the code appears on the screen and you've read it, you're already influenced.

Of course, nobody is forcing you to adopt the code, but the temptation is great. The learning effect would be greater if you developed more methods yourself. If you do fail with a problem, you can always ask for improvement tips for your half-finished code.

2.4.5   UML Diagram

The graphical summary in the form of a Unified Modeling Language (UML) diagram of classes, variables or attributes, or properties and methods is a valuable aid to maintaining an overview.

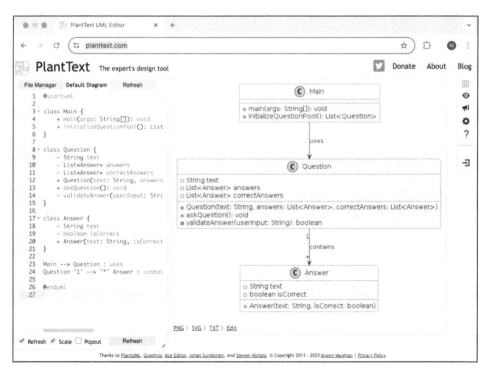

Figure 2.1 UML Diagram of the Quiz Program in a PlantUML Representation

As of the time of writing (early 2025), the current AI tools are unfortunately still hope-lessly overwhelmed in this respect. As the following prompt shows, not even the ASCII representation of the diagram is free of errors. (Note the right-hand edge of the boxes!)

Prompt: Can you present a UML diagram of the classes so far?

Certainly! Below is a UML diagram that represents the classes and their relationships as described so far for the quiz game. The diagram includes the Main, Question, and Answer classes with their respective attributes and methods.

```
+------------------------+
|         Main           |
+------------------------+
| + main(args: String[]): void |
| + initializeQuestionPool(): List<Question> |
+------------------------+


+------------------------+
|        Question        |
+------------------------+
| - text: String         |
| - answers: List<Answer> |
| - correctAnswers: List<Answer> |
+------------------------+
| + Question(text: String, answers: List<Answer>,
|            correctAnswers: List<Answer>) |
| + askQuestion(): void   |
| - validateAnswer(userInput: String): boolean |
+------------------------+
```

You can avoid this restriction by asking the AI tool in the prompt to provide code for a common diagram tool instead of the diagram. Mermaid (*https://mermaid.live*) and PlantUML (*www.planttext.com*) are well suited for this purpose. There are websites for these libraries where you can insert the UML code and display it immediately.

Prompt: Please generate PlantUML code instead.

```
@startuml
class Main {
    + main(args: String[]): void
    + initializeQuestionPool(): List<Question>
}
...
```

2.4.6   AI Bot or Code Wizard?

The more extensive your code becomes and the more work you delegate to AI tools, the less suitable it is to use a chat system such as ChatGPT or a similar tool. The problem is that the AI tool loses track of the code if you develop parts of it yourself.

Of course, you can insert the current status of the program into the prompt from time to time or make it available via drag and drop. You can avoid this increasingly tedious back and forth as the number of classes increases by using a code wizard that has permanent access to the entire code.

An argument against the use of a code wizard is that it not only becomes active when explicitly requested, but also makes suggestions for additions every time code is entered or changed. This is useful for writing code efficiently, but less so for learning.

2.5   Example: Solving Sudoku

In this final example, we'll show you how you can develop a Java program for solving sudoku games in conjunction with ChatGPT or another AI tool. The initial situation is that you're pursuing your own ideas, but you want to avoid tedious typing with the help of AI support. So, you give clear instructions as to which code the AI should generate, then incorporate this into your program, and finally develop it further yourself until you encounter problems. If you will, you're the "senior developer" who can delegate unpleasant work to a "junior AI developer".

2.5.1   Class Setup

Working step by step, first, you specify the structure of the program. You leave the AI bot with little choice. The prompt contains information about your knowledge, the preferred programming language, the planned program structure, and the data structure for storing the sudoku game.

> **Single-Prompt Sudoku Solver**
>
> At the end of the section, we'll show you that you can also get a sudoku solver with a single prompt, but that isn't your goal. As the saying goes, "It's the journey, not the destination," and in the course of this journey, you want to deepen your code structuring skills.

You explicitly ask the AI chat system to set up a Sudoku class with instance variables and a constructor, but without any other methods. In addition, the AI tool should implement the Main class and initialize a simple sudoku puzzle there as you need a sudoku puzzle as a starting point for the other tests.

> **Prompt:** I am an intermediate Java programmer. I want to write a program to solve
> sudokus. I want to do this step-by-step with your assistance. Please do NOT provide the
> complete code. Instead, help me to set up classes, add methods, and so on in small
> steps.
>
> My plan is to use (at least) two classes: Main with testing code and Sudoku to store and
> solve the sudoku. More classes might follow.
>
> The sudoku is to be stored in a 9 × 9 int array. 0 means this cell is empty.
>
> If this sounds OK, please provide the Sudoku class with a private instance variable for
> the array and a constructor to initialize the variable. The constructor should also verify
> the array size and its contents (only values between 0 and 9 allowed). Add no other
> code to the class.
>
> Also provide the Main class, which creates a Sudoku object and passes a simple but
> incomplete sudoku to solve.

You copy the two classes into a new Java project and have a functional basic framework
for your code in two minutes. The code for the Sudoku class looks as follows:

```
import ...;
public class Sudoku {
    private int[][] grid;  // store Sudoku cells in array

    // constructor
    public Sudoku(int[][] grid) {
        // Check if the input grid is valid
        if (grid.length != 9 || grid[0].length != 9) {
            throw new IllegalArgumentException(
                "Grid must be a 9x9 array.");
        }

        for (int i = 0; i < 9; i++) {
            for (int j = 0; j < 9; j++) {
                if (grid[i][j] < 0 || grid[i][j] > 9) {
                    throw new IllegalArgumentException(
                        "Grid values must be between 0 and 9.");
                }
            }
        }

        // Initialize the grid
        this.grid = new int[9][9];
        for (int i = 0; i < 9; i++) {
            for (int j = 0; j < 9; j++) {
                this.grid[i][j] = grid[i][j];
```

```
        }
      }
    }
}
```

2.5.2 Detaching Individual Fields in Rows

Now you can proceed as you would to solve a sudoku: in a first method, your program should run through all rows. If there is only one free field in a row, this can be filled with the only number that doesn't yet appear in the row.

In the prompt, you can formulate a loop over all rows as the first intermediate goal, but for the time being, only ask for suggestions for a meaningful name for the method.

> **Prompt:** Perfect. Now we start the solving code. As a first step, I want to loop through the grid to find rows with exactly *one* empty cell. It should be easy to insert the missing value into these positions. I am unsure how to name this method. Do you have any ideas?

Needless to say, ChatGPT does have some ideas for the name; it provides four to choose from. The AI tool also provides an unsolicited code suggestion using the method name, `fillSingleEmptyInRow`.

You look at the code in the chat system and are basically satisfied. Before you transfer the code to your program, however, you still have two requests for improvement, which you express in the following two prompts. First, more meaningful variables should be used in the code (ChatGPT initially opted for i and j), and second, the method should integrate a counter and at the end return how many fields were solved when running through the loop.

To prevent the AI tool from outputting the entire code of the class at each iteration, ask it to display only the relevant method. Notice how the AI tool understands from the context that the two prompts refer to the previous question.

> **Prompt:** Please replace i and j with `row` and `column` to make the code more clear.
>
> **Prompt:** Add a counter to `fillSingleEmptyInRow`. The method should return the number of cells it has filled. Only output the changed method, not the entire class.

The method programmed entirely by the AI tool now reads as follows:

```
public int fillSingleEmptyInRow() {
    int cellsFilled = 0;
    for (int row = 0; row < 9; row++) {
        int emptyCount = 0;
```

```
        int emptyColumn = -1;

        for (int column = 0; column < 9; column++) {
            if (grid[row][column] == 0) {
                emptyCount++;
                emptyColumn = column;
            }
        }
        if (emptyCount == 1) {
            boolean[] present = new boolean[10];
            for (int column = 0; column < 9; column++) {
                present[grid[row][column]] = true;
            }
            for (int number = 1; number <= 9; number++) {
                if (!present[number]) {
                    grid[row][emptyColumn] = number;
                    cellsFilled++;
                    break;
                }
            }
        }
    }
    return cellsFilled;
}
```

[explanation of changes]

2.5.3 Solving Individual Fields in Columns and in Blocks

Now you incorporate the method suggested by the AI tool into your program and test it. Once everything works to your satisfaction, you ask ChatGPT to write two equivalent methods for columns and for 3 × 3 blocks.

Because you've made a few minor changes to the class in the meantime, you transfer the entire code of the class in its current state to the prompt so that the AI tool has the correct starting point. (You can avoid this step if you use an AI wizard integrated in the editor or in the development environment.)

> **Prompt:** I reorganized the code a bit. This is what the class looks like now. Please provide two more methods similar to fillSingleEmptyInRow to handle columns with only one cell missing and 3 × 3 areas with only one cell missing.
>
> [code of the entire Sudoku class at this point in time, approximately 50 lines]

The code of the two new methods looks very similar to `fillSingleEmptyInRow`. You integrate the two methods into your code.

2.5.4   Test Code and Issues with the Test Data

To facilitate testing and debugging, ask ChatGPT to program two more methods. One should count the number of unsolved fields, while the other is supposed to display the sudoku in a legible form.

> **Prompt:** Please provide code for two more methods for the Sudoku class: one is to count the empty cells, and the other one prints the sudoku with empty spaces instead of 0.

You integrate the largely trivial code of these methods into your project. We haven't printed the code here. If necessary, take a look at the sample files for the book.

The code previously supplied by ChatGPT didn't contain any errors. However, there are some issues with the test data. The sample sudoku generated by ChatGPT at the beginning of the session is too difficult for your program in its current configuration. So, you need to ask for simpler test data. You explicitly point out in the prompt that you only need the code for the array (once unsolved, once solved). Otherwise ChatGPT will output the entire code of the Main class again.

> **Prompt:** I added calls to the two new methods to my solve method. However, the test puzzle you provided is too difficult at this time. Can you provide a slightly easier puzzle for testing? (Provide just the int[][] array and the solved puzzle for reference, no other code.)

The tests with the new sudoku are just as unsuccessful as with the original test data. You need a sudoku in which there is at least one row, one column, or one block in which only one number is missing. You formulate this information in another prompt.

> **Prompt:** Even easier please. We need at least one row, column, and block with only one cell missing to test the code so far.

Absurdly, ChatGPT also fails at the third attempt to deliver a sufficiently simple sudoku puzzle. You can also take a completely solved sudoku and replace various cells with 0 in the sense of "not completed" and continue your tests with it.

2.5.5   Brainstorming

This brings you to the point where your program works in principle, but only for very simple sudoku puzzles. Now you don't really know what to do. You ask ChatGPT for advice. For the AI system to get an idea of what your solution method currently looks like (you've written this code with calls to all other methods yourself), you want to pass this method to the prompt. Alternatively, you could add the file of the entire class to the prompt using drag and drop. In the prompt, you *do not* explicitly ask for new code, but for ideas on how you could further improve the program.

> **Prompt:** I added a loop to my solve method. It calls all three `fillSingleXxx` methods until they are unable to solve a single cell.
>
> ```java
> // in the Sudoku class
> public boolean solve() {
>     int n;
>     do {
>         n = 0;
>         n += fillSingleEmptyInRow();
>         n += fillSingleEmptyInColumn();
>         n += fillSingleEmptyInBlock();
>         System.out.println("Filled " + n + " cells.");
>     } while (n > 0);
>     if (countEmptyCells()==0) {
>         System.out.println("Sudoku puzzle solved!");
>         return true;
>     }
>     System.out.println("Could not solve the puzzle.");
>     printSudokuWithSpaces();
>     return false;
> }
> ```
>
> So far, my program can solve only very simple sudokus. Can you suggest a more sophisticated solution method? Just present the idea, no code please.

ChatGPT now presents common solution concepts for sudokus (*naked pairs/triples, hidden pairs/triples, X-wing,* etc.) on the one hand and proposes a recursive backtracking algorithm on the other. First, let the AI tool explain the concept of *naked pairs/triples*.

> **Prompt:** Let's start with naked pairs/triples. Can you show me a sudoku that can be solved with this concept so I can better understand the idea?

2.5.6   Storing Solution Candidates per Field

ChatGPT explains how the fact that only very few numbers are possible with a pair of free cells is exploited, thus further reducing the number of options. This gives you a new idea. Wouldn't it make sense to generally store for each cell which numbers are still permissible there? This information would speed up any further search for a solution.

You formulate this idea in the prompt. Your question only refers to a suitable data structure. Of course, you could simply ask ChatGPT for a suggestion. But you want to learn how to program on your own and are thinking about it yourself instead. The AI tool is more for verifying your ideas.

> **Prompt:** I want to add an instance variable with 9 × 9 sets to store candidates for each cell. I want each set to be initialized with a set of all numbers from 1 to 9. Is
>
> ```
> private Set<Integer>[][] candidates;
> ```
>
> the correct data type? Later, I will add a method to remove all numbers already used within the same row, column, or 3 × 3 block.

ChatGPT confirms that the data type is a good choice and also provides the code for the initial initialization without being asked. Each set is initialized with all nine numbers. You implement the changes and then commission the next method.

In the prompt, you explain what the method should do: run through all 81 fields of the sudoku and eliminate from the set of possible solutions for this field all digits that already occur in this row, column, or block. This leaves only the solution candidates. Programming this method manually wouldn't be particularly difficult, but it would be tedious and prone to errors due to the many nested loops.

> **Prompt:** Next, I need a method to loop over all the cells. For each cell, remove entries from the candidate set if the same number is used within the row, column, or 3 × 3 block. Can you provide code to do this?

You add the new `updateCandidates` method and call it after the three `fillSingleXxx` methods. In other words: First, the solution algorithm tries to find rows, columns, or 3 × 3 blocks with only one free cell. Second, `updateCandidates` eliminates all numbers already used in a row, column, or 3 × 3 block from the set of solution candidates. If there is now a cell for which there is only *one* possible solution, another field of the sudoku has been solved.

2.5.7   Filling in Fields Where Only One Solution Number Is Possible

In the next prompt, ask your AI junior programmer to program another method that searches for exactly such fields and enters the solution number there. As before, the method should count how many fields are solved in this way.

> **Prompt:** Write a method to loop over all the cells that are still empty. If the candidate set contains exactly one number, set that cell accordingly. Again, the method should return the number of cells solved.

You integrate the new method into the loop for solving sudoku puzzles, and lo and behold, the program can now solve even moderately difficult puzzles. At this point, you're satisfied with your work and the AI contributions, so you close the project for the time being. The Sudoku class now has almost 300 lines. (You'll find the complete code in the sample files for the book.)

2.5.8   Personal Conclusion

I programmed a sudoku solver in the Kotlin language a few years ago without the help of AI. It took me about two days to get to the state of the program outlined here. I enjoyed programming back then, especially researching different sudoku solving strategies. I then added one or two more methods and finally implemented a recursive search algorithm. The time required for this was approximately another two days (you can read more about this at *https://kofler.info/sudokus-mit-kotlin-loesen*, although you'll need to translate from German to English).

So, when I started this example, I already had some prior knowledge. Nevertheless, it was amazing that after *two hours* with the help of AI, I was about as far as I was on my first attempt after *two days*!

What I like best about this example is that even with this new implementation, I have the feeling that this is *my* code. Sure, thanks to AI help I have saved myself the tedious formulation of the nested loops. This isn't difficult, but it's prone to errors. But the AI-generated code is clearly understandable to me, and I could have changed it at any point or corrected it if necessary. (The latter wasn't necessary. The code generated by ChatGPT-4o was error free.)

Why did it work so well? Well, first, because there are various sudoku solvers available on the web, there was sufficient material at hand to train the language model. And secondly, because my own knowledge was relatively extensive, I set the direction of the code development, and I was able to ensure that the code made sense and that it was correct for each new method—almost at a glance.

This is exactly where I see the ideal conditions for the use of AI in coding: you know what you're doing and aren't overwhelmed by the complexity of the AI-generated code. You can easily assess whether the code works.

Admittedly, these prerequisites don't apply if you're just learning to program and solving typical exercises is indeed a major challenge. Ideally, you can also learn from the AI suggestions. However, you'll need a lot of self-discipline: your coding knowledge will only grow if you solve subproblems yourself first and question the code provided by the AI tool until you've understood it in detail.

2.5.9   Alternative Procedure

In this example, we've specified relatively clearly which strategies should be used to solve the sudoku. We've chosen a very "human" approach and used code to implement how we would solve a sudoku ourselves.

As the following prompt shows, one alternative is to leave the solution path entirely to the AI tool. However, this reduces your own contribution to the code to nothing.

> **Prompt:** Can you show me a short, concise way to solve a sudoku with Java?

Most AI tools then suggest the backtracking algorithm. This is how it works in principle: The numbers 1, 2, and 3 to 9 are inserted into each free cell in turn. As soon as a number doesn't cause any obvious errors (no duplicates in the row, column, or 3 × 3 block), the method continues this procedure for the next free cell. If none of the numbers 1 to 9 fits there, a mistake was obviously made earlier. At this point, the method then terminates, goes back to the previous field, and inserts the next number there.

The backtracking algorithm is recursive, that is, it's based on the fact that a method can call itself repeatedly. The recursive approach has the advantage that it's easy to return to an earlier stage of the solution search in the event of errors. This is where the name of this concept comes from.

However, this approach is completely "unhuman". Instead of starting with fields where there are obvious, simple solutions, the program systematically tries out all possibilities. The recursive approach is also very difficult to understand, especially if you're just learning to program. You can ask the AI tool to explain the code to you.

> **Prompt:** Great! However, it's hard for me to understand the code. Could you add some comments to the code and explain the relevant concepts?

Although the code with the backtracking algorithm is difficult to understand, it has two decisive advantages over the solution presented earlier:

- The code can be implemented in just a few lines. The entire SudokuSolver class (see the sample files) including main and a test sudoku takes up only 70 lines.
- The algorithm can cope with sudokus of any difficulty.

2.5.10   Prompting Correctly

This book isn't about solving programming issues, regardless of whether they are sudokus, REST APIs, or database applications. It's rather about learning to use AI tools efficiently and responsibly.

In this section, we've shown you how to solve a complex problem using the support of AI. As an alternative approach, however, we've also demonstrated that you may reach your goal much faster if you use the capabilities of AI tools without restraint. (This doesn't always work as well as when solving sudoku!)

We believe that the second way is the wrong one:

- If you're still learning, the first way will give you a basic understanding of how to work with arrays, structure the code, and so on. Even though many methods come from the AI tool, you can still add code yourself and learn a lot. This isn't the case with the alternative approach.

- Even if you're only interested in solving an issue quickly, AI-generated code that is incomprehensible to you isn't desirable. Ultimately, you're responsible for your code and must be able to explain, adapt, or correct it if necessary. No matter what level of IT knowledge you currently have, you should always use AI tools within the scope of your knowledge, not beyond it.

Chapter 3
Debugging

As soon as a software project exceeds a certain size, errors creep into the source code. The aim of debugging is to find and rectify these errors. The errors usually occur in the course of development, and users often don't point them out until later. However, we have to solve the problem, and if we can solve it faster by using AI, then all the better.

In software development, we have different tools to track down problems. Temporary `print` statements (or `console.log` for web applications) in the code are the simplest and often quickest way to success. Structured logging with different categories and other metadata are helpful in applications that are used in production by customers. A debugger that is able to analyze and stop the code at runtime is an indispensable tool for complex applications.

These tools help you recognize and understand errors, but they aren't magic remedies. It's still up to you to think through the problem, rule out sources of error, and investigate the cause. Often, you can get so engrossed in your own code that you don't even notice anything that doesn't make sense. Once you're stuck in a dead end, it's difficult to find a way out on your own.

This is why modern software development often involves teamwork: in *pair programming*, you work together on the code, because four eyes (and two brains) simply recognize problems more than twice as quickly. If no colleague is available, *rubber duck debugging* often helps: grab a rubber duck and explain the problem to it. Because the duck isn't a programming professional, these explanations have to be slow and in small steps. The solution is usually obvious once you've broken down the error in this way.

But what if the rubber duck wasn't just a passive listener, but could contribute with analyses, proposed solutions, and ideas of its own? This would give you a partner in pair programming who is never grumpy or bored, always has time, and also has a good basic knowledge of almost every language and framework.

In this chapter, we'll try to eliminate errors in our own source code with AI support. Thanks to our experience in fixing the errors, we can estimate how much time savings is realized using AI.

The disadvantage of this is that our knowledge of the solution path can affect the question. This is precisely the central point in the formulation of the prompt: if you already give your AI helper a direction and send it on a certain path using keywords, it will most likely pick up on your train of thought and replicate it. If this doesn't solve the problem, it may make sense to start a new session and formulate the request in a more open and general way. This contradicts the advice we usually give in the book, namely that the prompts should be formulated as concretely as possible and with a lot of context so that the AI tool knows what it has to do.

However, if you're not sure yourself where the problem is and what the right solution is, it's better to give the AI tool a little more leeway. Ask it for different approaches and perspectives on the problem. Then, of course, it's up to you to choose the path that solves the problem between the various ideas. AI can certainly help you debug simple syntax problems, but when it gets more complex, the bottom line applies all the more: always check all answers and keep them in mind. Just because your AI helper proudly says that it has solved a problem doesn't mean that everything is working as it should.

Nevertheless, we can attest that our experience with AI-supported debugging has been very positive. The language models are good at finding and correcting deviations from the norm. If the error in your program can be described in this way, there is a high probability that your AI helper will find it.

3.1 Web Applications

The market share of web applications continues to grow unabated. The COVID-19 pandemic has made the call for *more online* applications even louder, and anyone bringing an application online today would do well not to forget the web browser as a possible frontend.

As an operating system-independent platform, the browser is developing rapidly and offers increasingly sophisticated tools for debugging web applications (keyword: *developer tools*). We want to see whether and how AI support can lead us to even faster solutions.

3.1.1 CSS Fixes

Anyone who has ever tried to create a supposedly simple layout in HTML will be familiar with the following problem: an element simply doesn't want to be in the position on the page where you would like it to be. As an example, we've used a very simple page closure as a `footer` element. The content is supposed to be divided into three areas: left, center, and right. Our first design has the disadvantage that the areas with Cascading Style Sheets (CSS) property `float` don't adhere to the height of the surrounding element, but can also flow beyond it.

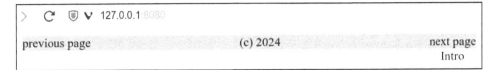

Figure 3.1 Incorrect Layout for the Footer of an HTML Page

Let's see if *Claude 3.5 Sonnet* can offer a better solution here:

Prompt: How can I make my footer elements all the same height:

```
<footer style="margin:auto;text-align:center;width:80%;">
  <span style="float:left;">previous page</span>
  <span>(c) 2024</span>
  <span style="float:right;">next page<br />Intro</span>
</footer>
```

To make the footer elements all the same height, you can use flexbox. Here's a modified version of your code that should achieve this:

```
<footer style="margin:auto; width:80%; display:flex;
    justify-content:space-between; align-items:stretch;">
  <span style="flex:1;padding:10px;">previous page</span>
  <span style="flex:1;padding:10px;text-align:center;">
    (c) 2024</span>
  <span style="flex:1;padding:10px;text-align:right;">next page
    <br />Intro</span>
</footer>
```

Key changes:

1. Removed float properties and used flexbox instead.
2. Added `display:flex` to the footer.
3. Used `justify-content:space-between` to distribute the spans evenly.
4. Used `align-items:stretch` to make all items stretch to the height of the tallest item. [more details]

The solution uses the powerful CSS Flexible Box Layout (Flexbox) model to position the elements correctly. Flexbox has been implemented in popular browsers for almost 10 years and can do much more than just divide an area into three equally wide columns. However, this can also make Flexbox very complex, which is why we want to ask the AI tool whether we can also use the more modern (and often simpler) grid layout.

Prompt: Would CSS `display:grid` be an option?

Yes, CSS Grid is definitely a good option for this layout. Grid can provide even more control and flexibility than Flexbox in many cases. Here's how you could implement the same layout using CSS Grid:

```
<footer style="margin:auto; width:80%; display:grid;
    grid-template-columns: 1fr 1fr 1fr;">
  <span style="padding:10px;">previous page</span>
  <span style="padding:10px;text-align:center;">(c) 2024</span>
  <span style="padding:10px;text-align:right;">next page<br>
    Intro</span>
</footer>
```

Key points about this Grid implementation: [more details]

The result is even easier to read, and the display is identical to the Flexbox version.

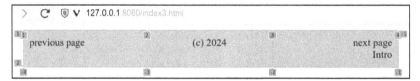

Figure 3.2 Grid Layout of an HTML Page Footer with Three Columns (with Active Developer Tools to Highlight the Three Areas)

For the first debugging task, the AI tool provided a very good solution to our problem. Even the variant we requested with a different CSS layout model remains clear and works straight away.

3.1.2 iCal Download in the Browser

Our next problem deals with the PHP programming language and the options of starting external programs from the browser. In our application, we've created an export for appointments in the standardized *iCal* format. However, clicking on the link doesn't start the calendar application as we wanted, but the text is displayed in the browser in a way that is difficult to read. We ask Claude 3.5 Sonnet if there is a quick solution:

Prompt: I have this simple PHP class that sends calendar entries over HTTPS. Unfortunately, it opens in my browser. How can I make it open in my calendar app?

To make the calendar entry open in your calendar app instead of the browser, you need to adjust the HTTP headers you're sending. Here's how you can modify your PHP class to achieve this:

[code response]

The key changes are in the `send()` method:

1. We set the `Content-Type` header to `text/calendar; charset=utf-8`. This tells the client that we're sending a calendar file.
2. We set the `Content-Disposition` header to `attachment; filename="event.ics"`. This tells the browser to treat the file as an attachment rather than trying to display it, and suggests a filename for the file.

The solution is correct, and the two changes, which are explained in detail as *key changes* in the answer, do indeed lead to the desired behavior. However, a mishap occurred during this attempt because we actually wanted to add the corresponding code passage to the prompt, so it was surprising when the AI-generated code integrated perfectly into our code base. For more complex changes, however, using your own code can certainly save further work.

3.1.3   Backend Frontend without a Proxy Server

If you use separate frontend and backend applications in web development, you'll probably be familiar with the following problem: the development server for the backend application runs on port 8000, and the bundler with live reload in the browser (e.g., Vite or Webpack) runs on port 8080. Asynchronous access from the frontend (Ajax) is blocked by the browser.

In such cases, the error messages in the browser console are detailed but difficult to understand if you haven't yet dealt with this issue. We've shortened the message in the following prompt that we sent to Claude:

Prompt: My backend application is a Golang JSON API with GIN-GONIC running on port 8000. Frontend is running with vite.js on port 8080. When I call backend routes, I get the following error:

```
Access to fetch at 'http://localhost:8000/users/me'
from origin 'http://localhost:8080' has been blocked by CORS
policy: Response to preflight request ...
```

What can I do?

To resolve this CORS (cross-origin resource sharing) issue, you need to configure your backend to allow requests from your frontend origin. Here's what you can do:

1. Install the CORS middleware for Gin:

[code response]

The AI tool recognized the problem immediately and solved it correctly. We need to include the cors module from the gin-gonic library we're using and configure it accordingly. We're also informed that the detailed configuration can be bypassed for development purposes if the config.AllowAllOrigins = true option gets set.

The problem seems to be solved—only the login still doesn't work. The request with user name and password is confirmed as valid with the HTTP status code 200, but we're getting constantly redirected back to the login page. Perhaps AI can help here too?

Prompt: Great! My requests are working now, but my login functionality is not. Session variables in Go aren't present. What is the problem?

I'm glad the CORS issue is resolved! Now, regarding your login functionality and session variables not being present, there could be a few reasons for this. Let's explore some common issues and solutions:

[more details]

The AI is obviously happy that we were able to solve the first problem. We're also pleased with the follow-up answer informing us that we need to make an additional setting at the frontend and backend when transferring access data. The AI tool couldn't have known this in advance, as we didn't say anything about a login issue when we first asked. The detailed answer also contains instructions on how to save the cookies used for the session correctly (so that they also apply to other ports).

The only thing we would have expected here is that the AI tool would immediately provide a suggested solution for the library we use (gin-gonic) and the session management there. The answer only explained the setting for cookies in general in the Go programming language. So, we have to ask again:

Prompt: I am using gin-gonic sessions, how do I set relaxed restrictions for SameSite here?

Now we get the correct answer, and the application works as desired. In the development process, a proxy server is often used for this purpose, which sends the frontend and backend to the browser from the same port and from the same host. If you want to run the application on a production system in a distributed manner (i.e., not on the same host), you'll have to deal with CORS anyway.

There is plenty of information about CORS available on the internet, and it's easy to find. A web search for the terms *CORS* (for *cross-origin resource sharing*) and *MDN* (for *Mozilla Developer Network*) leads us to excellent websites that describe the problem in detail. However, we don't get the finished solution for our program in the *Go* programming language, as provided by the AI tool. Of course, one could argue that this is a security-relevant topic and that code shouldn't be adopted without in-depth knowledge of the subject matter. In this case, however, AI provided good explanations, and we were able to develop the application in a very short time without spending half a day looking for solutions at Stack Overflow, which is what we did without AI help.

3.1.4 JavaScript Fetch with the Wrong Content Type

Our next problem stems from a change in which we replaced an old JavaScript library for executing asynchronous JavaScript (Ajax) web requests in the browser with the modern fetch interface. Most of the calls went smoothly, but the changeover was a little awkward in a few places. Will AI be able to provide a quick solution to this problem too?

For the experiment, we copy the entire function that triggers the problem and the error message from the browser console into the prompt:

Prompt: JavaScript: When I call this function:

```javascript
const _newComment = function (topic, meetingId, event) {
  event.target.disabled = true;
  fetch(`api/comments`, {
    body: JSON.stringify({
      id: topic,
      comment: this.newCommentContent.trim(),
      meetingId,
    }),
  })
    .then((json) => {
      this.comments.push({
        id: json.id,
        content: this.newCommentContent.trim(),
        date: moment().format("YYYY-MM-DD"),
        meeting: meetingId,
        user: this.$store("user").fullname,
      });
      this.newCommentContent = "";
    })
    .finally(() => {
      autosize.update(event.target);
      event.target.disabled = false;
      this.$focus.focus(event.target);
    });
};
```

I get this error message in my browser console window:

```
comments.js:8 Uncaught (in promise)
TypeError: Failed to execute 'fetch' on 'Window': Request with
  GET/HEAD method cannot have body.
    at Proxy._newComment (comments.js:8:3)
  ...
```

The error message you're seeing is due to the fact that you're not specifying the HTTP method in your fetch call, which defaults to GET. However, you're also trying to send a request body, which isn't allowed for GET requests. To fix this issue, you need to specify the HTTP method (likely POST) and include the necessary headers. Here's how you can modify your fetch call:

[response code]

Key changes:

1. Added method: 'POST' to specify that this is a POST request.
2. Added headers to specify the content type as JSON.
3. Changed then(json => { to then(response => response.json()).then(json => { to properly parse the JSON response.

Once again, the AI tool has impressed us: the error that the wrong HTTP method is used (GET instead of POST) when sending the data using fetch is almost obvious from the message in the browser console. But we didn't expect two other errors (the missing content type and the missing conversion of the response from text to a JavaScript object with response.json()) to be resolved straight away.

3.1.5   JamGPT

In our search for tools with AI support for debugging, we came across the *Jam* browser extension. You can use this extension to easily document errors that you notice when using a web application. You can use it to create and share screenshots or screencasts (i.e., a video of the screen recording). In addition—and this is the great strength of the extension—a lot of additional information (e.g., the output of the browser console or the network requests executed) is recorded and also saved.

All information gets uploaded to the *https://jam.dev* platform, where each case, or *jam*, can be viewed and commented on. The service is freely available after registration, and paid access is available as expected for business and enterprise customers. Among other things, these offer the connection to an existing ticket system and login via the company network.

Now you may rightly ask what this has to do with AI. In Jam's web interface, you'll find the **JamGPT** tab on the far right of the screen, and this is where AI comes into play.

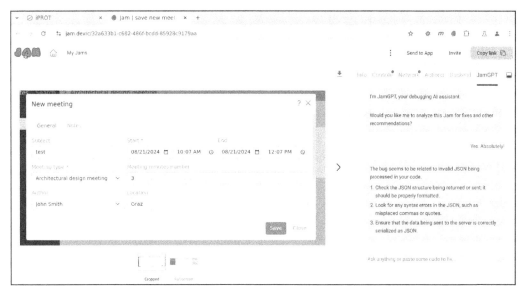

Figure 3.3 A Jam with AI Support for Faulty JavaScript Code

At the click of a button, you can send the error messages to a language model and hope for its solutions to the problem. The convenience of this solution is truly outstanding and demonstrates the possibilities, so we expected exceptional results. However, we were somewhat disappointed: for simple errors, such as a misspelled URL for a backend query, JamGPT correctly points out the possible errors, but a web developer doesn't need AI support for this. The error code 404 is well known.

However, the AI tool was unable to shine in a more complex error that we produced for JamGPT. On the contrary, the answer even pointed in the wrong direction: to validate form data, we used a library that analyzes the data-rules HTML attribute of an element and then checks the validity of the rules stored there. The value of data-rules must be a character string that can be converted into a JavaScript object using JSON.parse, for example:

```
<input type="text" data-rules='["required"]'  name="email" />
```

Fortunately, with current browsers, you can do without this type of JavaScript-based validation, but the code is already several years old. The preceding character string is converted by JSON.parse into a JavaScript array with one entry, namely the "required" string. If you forget the double quotation marks around the required keyword, an error occurs in the JavaScript code of the library. The JavaScript console provides information about the point at which the conversion fails:

115

```
Uncaught SyntaxError: Unexpected token 'r', "[required]" is not
  valid JSON
    at JSON.parse (<anonymous>)
    at validation.js:17:24
    ...
```

The response generated by JamGPT suggests that we check the server request or the server response for a valid JSON structure:

> The bug seems to be related to invalid JSON being processed in your code.
>
> 1. Check the JSON structure being returned or sent; it should be properly formatted.
> 2. Look for any syntax errors in the JSON, such as misplaced commas or quotes.
> 3. Ensure that the data being sent to the server is correctly serialized as JSON.

But we hadn't even gotten that far yet: the error occurs before we can send a request to the server. Admittedly, most errors when converting JSON into web applications occur during the communication with a server, which is why the AI tool also suggests this as a possible problem. In our case, however, the problem wasn't related to the server, and the context in the browser console would have shown this. We're curious to see whether Jam will make any further progress in the integration of language models, as the potential undoubtedly exists.

3.2   App Development

The following debugging examples refer to the Flutter library, which allows you to create apps for iOS and Android as well as web applications.

3.2.1   Flutter and Null Safety Values

The mobile app we created using the Flutter library was supposed to be multilingual. As a modern toolkit, Flutter naturally provides the tools for this, but we did run into a small problem: the application would no longer compile because the translation entries aren't *null safe*. Roughly speaking, the compiler can't ensure at compile time that the translation object is available at runtime.

The detailed error messages of the dart compiler already give a clear indication of what to do, but we also wanted to let the AI tool solve the problem:

Prompt: Flutter: My app isn't compiling after I inserted translations for labels.

```
ERROR: lib/widgets/team.dart:43:62: Error: Property 'warning'
  cannot be accessed on 'AppLocalizations?' because it is
  potentially null.
  - 'AppLocalizations' is from
  'package:flutter_gen/gen_l10n/app_localizations.dart'
  ('.dart_tool/flutter_gen/gen_l10n/app_localizations.dart').
  Try accessing using ?. instead.
  title: Text(AppLocalizations.of(context).warning),
                                            ^^^^^^^
```

This error occurs because the `AppLocalizations.of(context)` call might return null, and Dart's null safety feature is preventing you from accessing the warning property on a potentially null object. To fix this, you can use the null-aware operator `?.` as suggested in the error message. Here's how you can modify your code:

```
title: Text(AppLocalizations.of(context)?.warning ?? 'Warning')
```

As can be seen from the error message, the *null-aware operator*, that is, `?.`, should be used for the `AppLocalizations`. However, the AI tool goes one step further and appends the replacement text `'Warning'` (the double question mark provides the fallback for the null value in Flutter/Dart). Without this change, the app would still not compile because the surrounding `Text()` function expects a string, not `null`.

The AI tool also briefly explained the background to the changes, which we found very helpful:

This change does two things:

1. It uses the `?.` operator to safely access the warning property only if `AppLocalizations.of(context)` is not `null`.
2. It provides a fallback string ('Warning', in this case) using the `??` operator in case `AppLocalizations.of(context)` is null or `warning` isn't available.

Once again, we were pleasantly surprised by the AI tool's debugging capabilities.

3.2.2   Layouts in Flutter

App development using the Flutter library requires some rethinking in web development. If areas aren't completely filled by widgets or if they exceed the available space, Flutter fills the area with a black-and-yellow striped background to indicate the problem. This helps us to recognize the problem, but not to solve it, unfortunately. Let's see if AI can help us out here too:

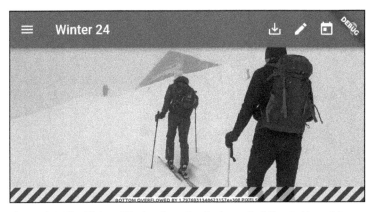

Figure 3.4 The Flutter Display for a Faulty Layout at the Bottom Edge

Prompt: Flutter: I get a yellow-black striped area at the bottom of my app. The widget I am using looks like this:

```
return Scaffold(
  drawer: const MyDrawer(),
  body: Column(
    children: <Widget>[
      SizedBox(
        width: MediaQuery.of(context).size.width,
        height: 250,
        child:
            Image.file(File(state.currentEntry.image),
              fit: BoxFit.cover),
      ),
      const TagsWidget(),
      MarkdownPreview(
        entry: state.currentEntry,
        picturePath: state.picturePath,
      ),
    ],
  ),
  appBar: AppBar(
      title: Text(state.currentTrip.title),
      actions: action.getActions(context)),
);
```

The yellow-black striped area at the bottom of your app typically indicates an overflow error in Flutter. This happens when a widget tries to render content that doesn't fit within its allocated space. In your case, the issue is likely caused by the Column widget in the Scaffold's body. [more details]

Claude 3.5 Sonnet explains two options in the following, both of which lead to the desired result. The generated code can be replaced without changes and it works.

The very brief explanation of the two different solutions (one with a `ListView` and one with a `SingleChildScrollView`) leaves us in the dark as to which is the better solution for our application. So, we ask the following:

Prompt: What is the difference between a `ListView` and a `SingleChildScrollView`?

`ListView` and `SingleChildScrollView` are both scrollable widgets in Flutter, but they have some key differences:

1. Child widgets:
   - `ListView` is designed to efficiently handle a potentially large number of children. It can have multiple child widgets.
   - `SingleChildScrollView` can only have a single child widget, typically a `Column` or `Row`.

2. Lazy loading:
   - `ListView` supports lazy loading. It only builds the child widgets that are currently visible on the screen, which is more memory-efficient for long lists.
   - `SingleChildScrollView` builds all its content at once, which can be less efficient for large amounts of content.

[more details]

Then follows additional points that succinctly describe the differences. Thus, AI doesn't disappoint us here either: even upon reading the first points, we realize that the second variant with the `SingleChildScrollView` is more suitable for our element, which is too large.

3.3 Developing on Raspberry Pi

When developing software on *embedded devices* such as Raspberry Pi, the development environment doesn't usually run on the device itself, but on the developer's workstation computer. This results in slightly different debugging requirements. To connect the debugger to the device, a network connection must be established with the corresponding port forwarding. If Raspberry Pi is running in a production system without internet access, good logging of the application is important to track down errors later.

We'll address these two topics in the following two sections.

3.3.1   Step-by-Step Debugging on a Raspberry Pi

Classic debugging involves setting *breakpoints* in the code and executing the program with a debugger. As soon as the point with the breakpoint is reached, the execution gets stopped, and it's possible to check the status of the application at this point. In addition, you can then follow each call *step by step* and examine the content of variables.

We haven't found a good application for AI for this activity. It remains manual work to set the breakpoints and check the application status. However, AI can support us with the setup so we can start step-by-step debugging in the first place. We've developed a Python application on a Raspberry Pi that controls a stepper motor and is controlled via an integrated touch display.

> **Prompt:** What tools do I need to debug my Python app step-by-step on a remote Raspberry Pi?

To step-by-step debug your Python app on a remote Raspberry Pi, you'll need the following tools:

1. SSH client: To connect to your Raspberry Pi remotely.
2. Remote debugging tool. There are a few options:
   - pdb (Python Debugger): Built-in Python debugger.
   - rpdb: Remote Python Debugger, an extension of pdb.
   - Visual Studio Code with Remote SSH extension: Provides a full IDE experience.
3. Code editor: If not using VS Code, you'll need a way to view and edit your code.
4. Port forwarding: To allow your local machine to connect to the debugger on the Pi.

Here's a basic setup using rpdb: [more details]

At the end of the instruction, which is not printed here, Claude 3.5 Sonnet mentions that the debugging process is smoother if you use VS Code and the Remote SSH plug-in. We want to know more about this:

> **Prompt:** How would I do it using VS Code with the Remote SSH extension?

Using Visual Studio Code with the Remote SSH extension is indeed a more user-friendly approach for remote debugging. Here's how to set it up:

1. Install Visual Studio Code on your local machine.
2. Install the "Remote - SSH" extension in VS Code.

[more details]

This is followed by 12 steps that briefly explain how to enable debugging for Python on the Raspberry Pi. Even if this sounds time consuming, you'll quickly arrive at a functioning debugging solution.

3.3.2   Help for Creating Log Entries

When debugging devices aren't on-site and are possibly offline, it's good practice to write important messages to a log file. This means that errors can also be discovered at a later date. We worked on a project in which the wireless network can be configured from a cloud platform on Raspberry Pi computers that are connected via the mobile network. A Python application on a Raspberry Pi makes the corresponding system calls to permanently save the settings.

We tried to get some support in creating the logging information from GitHub Copilot:

Prompt: Help me debug this function with some logging information.

```
def set_wifi_wpa_cli(ssid, psk):
  if ssid == '' or psk == '':
    return
  basecmd = ['sudo', 'wpa_cli', '-iwlan0']
  cmd = basecmd + ['list_networks']
  subprocess.run(cmd, capture_output=True)
  out_str = res.stdout.decode('utf-8')
  if re.search(rf"\t{ssid}\t", out_str, re.MULTILINE):
    # network already there
  else:
    cmd = basecmd + ['add_network']
    res = subprocess.run(cmd, capture_output=True)
    netnr = res.stdout.decode('utf-8')
    cmd = basecmd + ['set_network', netnr, 'ssid',
                     f'"{ssid}"']
    res = subprocess.run(cmd, capture_output=True)
    cmd = basecmd + ['set_network', netnr, 'psk', f'"{psk}"']
    res = subprocess.run(cmd, capture_output=True)
    ...
```

To add logging information to the `set_wifi_wpa_cli` function, you can use Python's built-in logging module. This will help you trace the execution flow and capture the output of each command.

```
logging.basicConfig(level=logging.DEBUG,
  format='%(asctime)s - %(levelname)s - %(message)s')
...
```

```
if ssid == '' or psk == '':
    logging.warning("SSID or PSK is empty.")
    return
...
if re.search(rf"\t{ssid}\t", out_str, re.MULTILINE):
    logging.info(f"Network {ssid} already exists.")
else:
    cmd = basecmd + ['add_network']
    logging.debug(f"Running command: {' '.join(cmd)}")
    res = subprocess.run(cmd, capture_output=True)
    netnr = res.stdout.decode('utf-8').strip()
    logging.debug(f"Network number: {netnr}")
...
```

As you can see in the excerpts from the response (we've abridged the output here), Copilot is doing a great job. First, logging is configured for Python. In a further step, the log outputs are analyzed according to their importance and marked accordingly with info, warning, or debug. This makes it easier for us to filter the log file later. For the system calls (subprocess.run), both the command and the result are logged. The AI also uses the relatively modern Python notation for character strings with *F-strings*.

3.4 Visual Studio and Visual Studio Code

The next two sections deal with two important development environments (*integrated development environments* [IDEs]) from Microsoft: Visual Studio and VS Code. The former can confidently be described as a cornerstone in the history of IDEs. The first version for Windows was launched back in 1997. Until 2014, the proprietary software was only available for a fee; since then, there has been a *Community Edition* that can be used free of charge after registering with Microsoft. The IDE is very popular because it offers context-sensitive help and a syntax check for many supported programming languages. The program requires a lot of resources from your computer, which is why many developers looked for leaner alternatives.

Since 2015, Microsoft has been developing the *VS Code* editor on the *Electron* platform, which was originally created as the basis for the free *Atom* editor. Electron provides a runtime environment for web applications that enables desktop applications without the limitations of the browser sandbox. The core of VS Code is lean and can be extended with countless extensions. In surveys on Stack Overflow, VS Code has been at the top of the list of developers' favorite IDEs for years.

3.4.1   Missing Python Modules in VS Code

The Python application in the following section was already a few years old when we had to deal with it again. An exception occurred when calling the debugger because an integrated module couldn't be found. Unfortunately, the application was created without a Python virtual environment and also without a `requirements.txt` file.

However, the AI support in VS Code based on GitHub Copilot and the corresponding extension were able to save us some work here too. The exception stops the execution and positions the cursor at the corresponding point in the code. The language support for Python in VS Code (again an extension) already indicates the nonexistent import. With Copilot, we can now have a solution suggested at the touch of a button (`/fix`) and insert the corresponding command directly into a console window. Then, simply press the Enter key to install the module.

Figure 3.5 GitHub Copilot Can Insert the Call for the Installation of the Missing Python Module Directly into a Terminal

Unfortunately, Copilot in VS Code can't yet handle the text of *exceptions* that occur at runtime. (We'll show how this is possible in Visual Studio in the following section.) However, we suspect that this function isn't too far away.

3.4.2   Visual Studio Exception Debugging

For users of Microsoft's Visual Studio development environment (not the free VS Code), who are also customers of GitHub Copilot, a very convenient AI extension is now available: if you start your application in debugging mode and an unexpected error (an *exception*) occurs, Visual Studio provides the option of having Copilot investigate the problem at the touch of a button.

Our experiments with this function left a very positive impression. The prompt not only sends the text of the exception but also the surrounding code. This was the only way we could understand why the (deliberately caused) System.ArgumentException could be explained so precisely. The C# call todoItems.GetRange(0, 3) extracts the first three elements from the todoItems list. However, the list was empty when we called it, which is why the exception occurred.

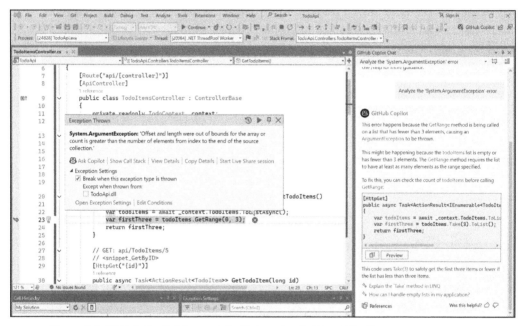

Figure 3.6 GitHub Copilot Support for Debugging in Visual Studio

GitHub Copilot was able to describe the problem briefly and with the correct variables:

This error happens because the GetRange method is being called on a list that has fewer than 3 elements, causing an ArgumentException to be thrown.

This might be happening because the todoItems list is empty or has fewer than 3 elements. [more details]

The fact that we also got a finished *fix*, that is, a code suggestion that works without any changes, was the icing on the cake. Microsoft is thus demonstrating the convenience of development with AI support. As I said, we're already looking forward to the implementation for VS Code.

3.5 Conclusion

If we conclude this chapter by asking ourselves whether AI support can be useful for debugging, we can undoubtedly answer *yes*. We don't see as much added value here as with code completion, but quick help for widespread problems, such as the CORS problem in Section 3.1.3, is a benefit.

We also expect the tools for AI integration to be further improved. What Microsoft is already demonstrating with the integration of GitHub Copilot in Visual Studio will probably soon be incorporated into other areas and products.

The area where AI support can once again show its strength is in programming languages that you don't use on a daily basis or are just learning. Small errors can lead to several hours of internet research on Stack Overflow and other forums, while AI can often not only describe the problem succinctly but also generate the corrected source code.

Chapter 4
Refactoring

The source code of an application isn't a static structure that is written once and then never touched again; rather, it's a dynamic entity that you have to deal with very often. In most cases, you read the source code and have to modify it much less frequently. This modification is often referred to as *refactoring*.

Refactoring is an important part of software development as it serves to improve the existing source code. There can be many reasons for refactoring, including cleaning up existing code that was created quickly, realizing that something has been learned in the course of developing the application or that application conventions have changed, making necessary modernizations because new versions of the libraries and frameworks have been released, and so on.

No matter what motivates you to refactor, it's always a critical and usually time-consuming intervention in your application. Luckily, tools are available to support you. Modern development environments can provide support for simple tasks such as renaming or extracting structures, but even they reach their limits at some point. This is precisely where AI tools can provide good support.

In most cases, refactoring isn't about creative and challenging tasks, but about adapting many code sections in the same way. AI tools really come into their own with routine tasks like these, as they can relieve you of repetitive tasks and thus reduce errors.

However, along with the advantages of using AI tools comes some risks. If you're aware of these risks, you can take countermeasures at an early stage to minimize them. In this chapter, you'll find out what these potential problems are and how you can deal with them.

4.1 Introduction to Refactoring

Before we dive into refactoring using AI tools, let's first take a look behind the scenes and find out exactly what refactoring is all about so that we can get the best possible help. *Refactoring* is the process of improving the internal structure of an application's source code, so you should separate the development of new features from refactoring. With refactoring, the focus is instead purely on the existing internal structure. The interfaces to other parts of the application or even to other systems remain the same.

4.1.1   Objectives of Refactoring

Refactoring should never be an end in itself. Instead, you should plan what you specifically want to improve. If you don't adhere to this, you can change the code, but you won't improve it. The aim of refactoring is to address one or more of the following aspects:

- **Improved readability**
  The structure of the source code gets designed in such a way that the code becomes easier to read and understand.

- **Increased maintainability**
  Refactoring adapts the structure of the code to make future changes and extensions easier.

- **Reduction of errors**
  Known sources of error can be minimized through targeted refactoring.

- **Increased efficiency**
  With source code that has grown over a longer period of time, inefficient structures creep in that slow down execution. Refactoring can improve these areas and thus increase the performance of the application.

Through continuous refactoring, you create a stable and flexible code base that enables you to react faster and more reliably to new requirements. Your application therefore remains future proof in the long term.

4.2   Refactoring Using AI Tools

Now that we've briefly dealt with the theory of refactoring, it's time to work with the source code. First of all, you'll learn how to assess whether you should revise a piece of source code at all, the criteria for making this decision, and how AI tools can help you. You'll also learn some strategies and best practices for specific refactoring processes.

4.2.1   Recognizing the Need for Refactoring

Your time as a developer is very valuable, so you shouldn't waste it on unnecessary work. It's therefore important to first find out whether refactoring is expedient and where the weaknesses in the code can be found.

There are numerous tools available for this type of analysis, such as SonarQube, which perform a static code analysis. These tools can provide a good overview of the overall system. This is also the weakness of many AI tools, as they rarely have an overview of the entire context of your application, but only smaller parts. A combination of different tools is therefore ideal: classic static code analysis for the overview and AI tools for the details and the actual refactoring. You can also use the AI tools as a kind of code reviewer that uncovers refactoring potential.

Code Smells

You can take the term *code smells* almost literally. Code that has code smells harbors potential problems that you can fix with refactoring. Code smells aren't necessarily bugs, but if you ignore them for too long, they can cause serious problems. Following are typical code smells:

- **Duplicate code**
  The code has the same or similar blocks in different places. If you correct an error at one point, you must also do this at all other points. Maintainability is therefore reduced, and the potential for errors is high.

- **Long methods**
  Functions and methods should be as compact as possible and only deal with one aspect. The more lines a function has, the greater the potential complexity.

- **Large classes**
  The same applies to classes as to functions and methods. If a class is too large, measured by the number of lines, properties, or methods, there is a high probability that it will do too much.

- **Feature envy**
  This type of code smell refers to a method that accesses more properties of another class than its own. This is an indication that the method is wrong at this point.

- **Data clumps**
  These are groups of variables that are used in the same way at different places in an application. They should be combined in a separate class or data structure.

Code smells are present in various forms in almost every application, but if you use the right tools, you can easily detect and rectify them. For example, we use a TypeScript class named OrderProcessor, which reads as follows:

```
class OrderProcessor {
  orders: any[] = [];

  addOrder(order: any) {
    this.orders.push(order);
    console.log("Order added:", order);
    if (this.orders.length > 100) {
      console.log("Too many orders");
    }
  }
}
```

```
processOrders() {
  for (let i = 0; i < this.orders.length; i++) {
    let order = this.orders[i];
    console.log("Processing order:", order);

    if (order.status === "NEW") {
      console.log("Order is new");
    } else if (order.status === "PROCESSING") {
      console.log("Order is being processed");
    } else if (order.status === "COMPLETED") {
      console.log("Order is completed");
    } else {
      console.log("Unknown order status");
    }

    if (order.customer.address.city === "Berlin") {
      console.log("Customer is from Berlin");
    }
  }
}

generateReport() {
  let report = "Order Report\n";
  report += "=============\n";

  for (let i = 0; i < this.orders.length; i++) {
    let order = this.orders[i];
    report += `Order ${i + 1}: ${order.id}\n`;

    if (order.customer.address.city === "Berlin") {
      report += "Customer is from Berlin\n";
    }
  }

  if (this.orders.length > 100) {
    report += "Warning: Too many orders\n";
  }

  console.log(report);
}
}
```

The class has the orders property in which it saves orders. The addOrder method adds new orders, the processOrder method processes the orders, and the generateReport method generates a report. You can use almost any AI tool for analysis. For example, you could copy the code into the Gemini console and ask for an analysis.

In this example, we use GitHub Copilot and Visual Studio Code (VS Code) integration. Here you can use the combination of the chat feature and workspace to analyze the code. To do this, open the class in the development environment, and enter the following prompt in the chat window:

Prompt: @workspace /explain Does the OrderProcessor class contain code smells?

Yes, the OrderProcessor class contains several code smells. Let's go through them step by step:

[explanation of code smells in the code]

Here is a refactored version of the OrderProcessor class addressing these issues:

```
interface Order {
  id: number;
  status: 'NEW' | 'PROCESSING' | 'COMPLETED';
  customer: {
    address: {
      city: string;
    };
  };
}

class OrderProcessor {
  ... updated code ...
}
```

Changes made:

[explanation of changes made in the code]

These changes improve the readability, maintainability, and robustness of the code.

This prompt is kept very general so that you give the tool plenty of scope to analyze the code and find code smells. Regardless of which tool you use, it's important with such a prompt that you make the code available as comprehensively as possible and not just individual sections. The more context you give the AI tool, the better it can do its job.

In the prompt, you explicitly asked for code smells. This is an established term that is sufficiently described in the literature so that AI can work with it. The use of such technical terms, such as certain design patterns, can make the prompts much more compact, as you don't have to describe the problem in a complex way. Try to use technical terms in your prompts so that you describe what you want to achieve as precisely as possible.

GitHub Copilot has found some vulnerabilities in our code. The tool identifies the code smells, explains what the problem is, and also provides an improved version of the class, including a detailed explanation of the changes. The following code smells were found and fixed:

- **Use of the any type**
  If you use the any type in TypeScript, you lose the advantages of the static type system. GitHub Copilot has created the order interface for the order and uses it throughout the methods and properties of the class.

- **Use of magic numbers**
  You should avoid storing numbers with a specific meaning, such as the limit for orders, directly in the code. Copilot provides a better solution here by storing the number in the form of the MAX_ORDERS property.

- **Repetitions in the code**
  The check as to whether the city in the order is Berlin can be found twice in the code and can be outsourced.

- **Long methods**
  The processOrders and generateReport methods are relatively long. Copilot has simplified the iteration of the orders by using the forEach method and has also outsourced parts of the methods so they are more compact and easier to read.

- **Missing error handling**
  When accessing properties such as city in the order city check, the code assumes that all properties in the chain exist. If that isn't the case, the JavaScript engine will throw an exception. For this reason, you should install an error handling routine here.

- **Console outputs**
  Logging using the console.log method isn't always the best option. Ideally, you should use your own logger for this, whose behavior you can better control.

The information provided by GitHub Copilot is very good, and the generated result is also significantly better than the original version of the source code. However, some points of criticism aren't taken into account (or only partially so) in the result, so you still have some work to do. The duplicate code was only moved to the checkCustomer-City method in one place. There is also no solution for using the logger, although the application-wide context is also missing here.

One of the code smells that Copilot didn't find is that the class covers too many responsibilities. Managing orders and generating a report don't necessarily have anything to do with each other and could be outsourced to separate classes. Access to the properties of the order address is an example of feature envy. The OrderProcessor shouldn't access the address object directly. Instead, it's better to introduce a separate class that provides you with the corresponding methods.

You can also be much more specific if, for example, you have a class or function that you're not happy with but don't know exactly how to modify it. The following function contains unnecessary duplicates in the code that the AI tool should resolve:

```
function processOrders(orders) {
  let totalRevenue = 0;
  let totalDiscountedRevenue = 0;
  let highValueOrders = [];
  let discountedOrders = [];

  for (let i = 0; i < orders.length; i++) {
    const order = orders[i];

    if (order.status === "completed") {
      totalRevenue += order.price;
    }
    if (order.status === "completed" && order.discount > 0) {
      totalDiscountedRevenue += order.price - order.discount;
      discountedOrders.push(order);
    }
    if (order.status === "completed" && order.price > 100) {
      if (order.discount > 0) {
        discountedOrders.push(order);
      }
      highValueOrders.push(order);
    }
  }

  return {
    totalRevenue,
    totalDiscountedRevenue,
    highValueOrders,
    discountedOrders,
  };
}
```

In this code example, you check whether the status of the order is completed in each if block. Such code is created quickly if you translate the requirements directly into code and don't make any optimizations. However, ChatGPT or another tool can do this for us with the following prompt:

Prompt: Could you please reduce the code duplicates in the following function:

[function code]

You can reduce the code duplication in the processOrders function by combining the conditions that check for the order's status and applying the relevant logic inside a single block. Here's a refactored version of your function:

```
function processOrders(orders) {
  let totalRevenue = 0;
  let totalDiscountedRevenue = 0;
  let highValueOrders = [];
  let discountedOrders = [];

  for (const order of orders) {
    if (order.status !== "completed") continue;

    totalRevenue += order.price;

    if (order.discount > 0) {
      totalDiscountedRevenue += order.price - order.discount;
      discountedOrders.push(order);
    }

    if (order.price > 100) {
      highValueOrders.push(order);
    }
  }
  return {
    totalRevenue,
    totalDiscountedRevenue,
    highValueOrders,
    discountedOrders,
  };
}
```

Changes made: [explanation of the changes made]

The AI tool has rewritten the code so that the multiple checks of the order status have been summarized. If the status doesn't correspond to the "completed" value, the system continues directly with the next run. This makes the code more compact and easier to read.

So, if you're not satisfied with your source code and know what changes you want to see, you can formulate this in a prompt, and the AI tool will adapt your code accordingly. Of course, there's also code that can't be optimized so easily. In this case, you should check the adjustments very carefully, as AI tools tend to make changes that don't really improve the code.

Anti-Patterns

You've probably heard of *design patterns*. These are established and reusable solutions for frequently occurring problems in software development. *Anti-patterns* are, to a certain extent, the opposite. They're often patterns or practices in the source code that cause problems in the long term. These solutions are usually the result of good intentions, but turn out to be problematic during further development. Typical anti-patterns include the following, for example:

- **God object**
  The god object or the god class is, in a way, the further development of a large class. It's a central component of an application. Over the course of development, more and more aspects have been added to the class, so that it takes care of too many aspects. This makes the source code much more difficult to maintain.

- **Incorrect use of the singleton pattern**
  In general, the singleton pattern has a bad reputation because it can easily be used incorrectly. For example, it can be used to manage the global status of an application.

- **Spaghetti code**
  If source code has a poor or inadequate structure, you can compare it to a mountain of spaghetti: you pull on one end and the whole mountain moves. This kind of source code makes it very difficult to correct errors or develop features any further.

- **Golden hammer**
  This anti-pattern refers to the fact that an established solution is used for all possible problems, even if it's not suitable for them.

- **Premature optimization**
  In applications, you often encounter the problem that the code is overengineered; that is, the solutions are unnecessarily complicated and solve problems that haven't even arisen yet.

The `UserService` TypeScript class serves as the basis for the analysis and subsequent improvement:

```
class UserService {
  private static instance: UserService;

  private constructor() {}

  static getInstance(): UserService {
    if (!UserService.instance) {
      UserService.instance = new UserService();
    }
    return UserService.instance;
  }
}
```

```
processUserData(users: any[]): void {
  users.forEach((user) => {
    if (user.isActive) {
      console.log(`Processing active user: ${user.name}`);
      if (user.role === "admin") {
        console.log("Admin privileges granted.");
        if (
          user.lastLogin >
          new Date().getTime() - 1000 * 60 * 60 * 24
        ) {
          console.log("Recent login detected.");
        } else {
          console.log("Login outdated.");
        }
      } else if (user.role === "user") {
        console.log("Standard user.");
        user.permissions.forEach((perm: any) => {
          if (perm.isEnabled) {
            console.log(`Enabled permission: ${perm.name}`);
          }
        });
      } else {
        console.log("Unknown role.");
      }
    } else {
      console.log(`Inactive user: ${user.name}`);
    }
  });
}

notifyUsers(users: any[]): void {
  this.processUserData(users);
}

calculateDiscounts(users: any[]): void {
  const discountCache: { [key: string]: number } = {};

  users.forEach((user) => {
    if (discountCache[user.id]) {
      console.log(
        `Cached discount for user ${user.name}: ${
          discountCache[user.id]
        }`
      );
```

```
    } else {
      const discount = user.purchaseHistory.length * 0.05;
      discountCache[user.id] = discount;
      console.log(
        `Calculated discount for user ${user.name}:
          ${discount}`
      );
    }
  });
}

logUserActivity(user: any): void {
  console.log(`Logging activity for user: ${user.name}`);
}

generateUserReport(users: any[], reportType: string): void {
  console.log(
    `Generating ${reportType} report for
      ${users.length} users.`
  );
}
}
```

As in the preceding example, we use GitHub Copilot in VS Code for refactoring with regard to anti-patterns. You can use the Copilot chat integration to submit the following prompt to analyze the code of the class:

Prompt: @workspace /explain Does the code of the UserService class contain anti-patterns?

Yes, the UserService class contains several anti-patterns and areas for improvement. Let's go through them step by step.

[explanation of anti-patterns in the code]

Refactored example:

Here's a refactored version addressing some of these issues:

```
interface User {
  id: string;
  name: string;
  isActive: boolean;
  role: 'admin' | 'user' | 'unknown';
  lastLogin: number;
  permissions: Permission[];
```

```
    purchaseHistory: any[];
}

interface Permission {
  name: string;
  isEnabled: boolean;
}

class UserService {
  private static instance: UserService;

  private constructor() {}

  static getInstance(): UserService {
    if (!UserService.instance) {
      UserService.instance = new UserService();
    }
    return UserService.instance;
  }

  processUserData(users: User[]): void {
    users.forEach((user) => {
      if (user.isActive) {
        this.processActiveUser(user);
      } else {
        console.log(`Inactive user: ${user.name}`);
      }
    });
  }

  private processActiveUser(user: User): void {
    console.log(`Processing active user: ${user.name}`);
    switch (user.role) {
      case 'admin':
        this.processAdminUser(user);
        break;
      case 'user':
        this.processStandardUser(user);
        break;
      default:
        console.log("Unknown role.");
    }
  }
```

```
  private processAdminUser(user: User): void {
    console.log("Admin privileges granted.");
    const oneDayInMillis = 1000 * 60 * 60 * 24;
    if (user.lastLogin > new Date().getTime() - oneDayInMillis) {
      console.log("Recent login detected.");
    } else {
      console.log("Login outdated.");
    }
  }

  private processStandardUser(user: User): void {
    console.log("Standard user.");
    user.permissions.forEach((perm) => {
      if (perm.isEnabled) {
        console.log(`Enabled permission: ${perm.name}`);
      }
    });
  }

  notifyUsers(users: User[]): void {
    this.processUserData(users);
  }
  calculateDiscounts(users: User[]): void {
    const discountCache: { [key: string]: number } = {};

    users.forEach((user) => {
      if (discountCache[user.id]) {
        console.log(`Cached discount for user ${user.name}:
${discountCache[user.id]}`);
      } else {
        const discount = user.purchaseHistory.length * 0.05;
        discountCache[user.id] = discount;
        console.log(`Calculated discount for user ${user.name}:
${discount}`);
      }
    });
  }

  logUserActivity(user: User): void {
    console.log(`Logging activity for user: ${user.name}`);
  }

  generateUserReport(users: User[], reportType: string): void {
    console.log(`Generating ${reportType} report for ${users.length}
```

```
users.`);
  }
}
```
Summary: . . . By addressing these points, the code becomes more maintainable, readable, and testable.

Once again, GitHub Copilot has done a great job, finding some weaknesses in the code and suggesting an alternative implementation:

- **Singleton pattern**
  Copilot has recognized the singleton pattern in the code and points out that you should be careful when using it.

- **Type safety**
  The original code used the `any` type. Copilot has created a `User` interface and uses it throughout the class.

- **Single responsibility principle**
  The `processUserData` method performs too many tasks. In its alternative implementation, Copilot has neatly outsourced the responsibilities to its own methods.

- **Reduce nesting**
  When nesting, the `processUserData` method is again noticeable in a negative way. By splitting the method, Copilot has also solved this problem. The combination of the violation of the single responsibility principle and the poor structuring indicates that we're dealing with spaghetti code.

- **Magic numbers**
  The calculation of the number of milliseconds in a day isn't obvious at first glance. In the revised version of the code, this calculation is stored in an appropriately named variable.

- **Logging**
  In its analysis, Copilot notices that `console.log` is used in the class and suggests using a separate logger instead.

- **Error handling**
  Another point of criticism in the analysis is the lack of error handling.

As with the analysis of the code smells, Copilot gives us a mixture of a concrete alternative implementation and some hints for the anti-patterns that we have to take care of ourselves. You can use the result of this prompt in your application without hesitation, as it significantly improves the existing code in any case.

The original code has several other vulnerabilities that haven't been addressed:

- **Too many responsibilities**
  The class covers too many responsibilities at once. The `logUserActivity`, `generateUserReport`, or `calculateDiscounts` methods provide information on this.

- **Premature optimization**
  The cache in the `calculateDiscounts` method could be an indication of an unnecessary optimization that increases the complexity of the code. At the very least, it would be good to have a corresponding note at such a point.

- **Inappropriate use of a method**
  The `notifyUsers` method is merely an alias for `processUserData`. Such code sections can indicate an anti-pattern.

When working with existing code, there are some best practices that can help you avoid problems. You can find out how to proceed in the following sections.

4.3   Best Practices

During refactoring, you may find yourself losing the overview and straying from your actual plan because you keep finding new places in the code that you want to improve. Here are some best practices to help you focus on the essentials.

4.3.1   Only the Internal Structure May Be Changed

When refactoring, you should concentrate on modifying only the internal structure of a function, but not its parameter list or return value. If you ask an AI tool for a refactoring, it will usually adhere to this recommendation. To be on the safe side, you can also specify in your prompt that the signature of the method shouldn't be changed.

> **Prompt:** Don't change the signature of the method; only focus on the inner structure.

This ensures that the refactoring doesn't affect the rest of your application and takes effect only in separate individual places.

4.3.2   Small-Step Approach

Even if you only focus on the internal structure of a method or class, refactoring can take on larger proportions, depending on the role the respective structure plays in your application. To prevent refactoring from getting out of hand, you should try to proceed in as small steps as possible.

If you use AI tools such as GitHub Copilot or ChatGPT, you should first analyze your code and see what refactoring potential the class or function has. Then, you can address the individual points step by step. This procedure has the advantage that you can commit the individual steps to the version system. Thus, if an error occurs, you're at most one step away from a functioning implementation and can repeat the last step using a modified prompt.

If the refactoring steps are too large and too many aspects are refactored at once, the risk of errors and undesirable behavior is significantly greater.

4.3.3   Test-Driven Refactoring

Before starting a refactoring process, you want to make sure that the class or function you intend to improve has been validated by automated tests. In most cases, these are unit tests or integration tests that check the interfaces of the subsystem you want to improve. If the tests still run successfully after the change, this is a first indication of successful refactoring. However, if the tests fail, you must reverse the change, check it, or at least analyze why the tests failed.

Automated tests don't replace a manual check of the refactoring result or a manual test, but they do give you security and relieve you of some of the manual testing effort.

4.3.4   Defining Clear Goals

Especially when refactoring using AI tools, there is a great risk that things will get out of hand and you'll rebuild too much. Before you start adapting the source code, you should set yourself a specific goal, such as improving the readability of the source code or introducing certain structural improvements. Prior to each refactoring step, you should check whether the desired modification will move you closer to your goal. If that isn't the case, you shouldn't do it.

In this chapter, you've seen that AI tools such as GitHub Copilot provide fast and comprehensive refactoring procedures. To really work in small steps and in a goal-oriented manner, you should limit the result with prompts that are as precise as possible.

4.4   Conclusion

AI tools don't replace manual refactoring, but they are invaluable aids that can relieve you of work and support you, for example, in not only analyzing the source code but also implementing a refactoring process. A key advantage of these AI tools is their ability to see patterns in code that can be difficult for the human eye to recognize. They analyze the source code in a depth and speed that can't be achieved by a developer alone.

Not only do these tools provide concrete suggestions for improvements, they often also explain the background to their recommendations. This allows developers to refactor code more efficiently and expand their own expertise and understanding of best practices in refactoring. Through these explanations and tips, AI tools promote a deeper understanding of the underlying problems and the proposed solutions.

Another significant advantage is the consistency guaranteed by the use of AI tools. While human developers can make mistakes due to fatigue or time pressure, AI algorithms work tirelessly and without subjective influences. This ensures that refactoring is carried out consistently and according to best practices.

However, it's important to note that AI tools need to be monitored, as they can occasionally overshoot the mark and introduce errors through hallucinations. These hallucinations can lead to the AI tool suggesting changes that don't correspond to the actual requirements or the desired behavior of the system. It's therefore crucial that human developers carefully review and validate the changes suggested by AI.

In summary, AI-supported tools are a valuable addition to manual refactoring. They improve the efficiency and quality of the refactoring process and allow developers to focus on more creative and challenging tasks. Despite these benefits, it's important to emphasize that human expertise and critical thinking are still essential to fully understand and consider the context and specific requirements of a project.

4

Chapter 5
Testing Software

Test tools and frameworks have become established in software development and are standard when it comes to quality assurance. Testing encompasses far more than just classic unit testing. It starts with the creation of meaningful test data for development and manual tests, spans through the validation of individual code blocks, and extends to end-to-end testing. Each of these test types poses its own challenges and requires specific approaches and tools. You can access various AI tools in each area and have them do a lot of work for you.

However, software tests have one thing in common: they are often repeated, as a basic set of test patterns is used again and again. This is why writing tests isn't a favorite task of software developers and often has the reputation of taking up an unnecessary amount of time and tying up resources that could be used for feature development. This is precisely where AI tools can play a decisive role by efficiently taking over repetitive tasks and significantly reducing the testing effort.

The use of AI tools isn't limited to a specific level of testing, but rather can be used as an aid anywhere. With the use of correctly formulated prompts and the right context, the AI tools can cover many test cases automatically. This support not only reduces the time required but can also recognize test cases that developers may not have thought of. In addition, AI can provide a "second opinion" or other perspectives, which makes the tests better and more robust overall.

In test-driven and test-accompanying development, AI can make a decisive contribution to test automation and ensure comprehensive test coverage. Yet, despite the many advantages, there are also challenges. The AI tool isn't always right and can sometimes generate false information through hallucinations. It's therefore important that you always check the tests generated by AI and at least subject them to a plausibility check. After all, it's about securing the software, and you can't take any risks here.

In the future, AI will play an even greater role in software testing. It can take a lot of work off your hands and allow you to focus on the actual development work. Nevertheless, AI remains a tool that—like any other tool—must be used and monitored correctly. The combination of human expertise and the efficiency of AI promises to take software testing to a new level and sustainably improve the quality of software products.

5.1   Generating Test Data

During development, you always have to deal with data, whether it's a form for registering new users, a list display of customer data, or a validation function for an international bank account number (IBAN). Ideally, you should already be using realistic data during development. This is exactly where the problems begin. What should realistic data look like, and where does the data come from? A concrete example of this case is the implementation of the customer list, which is frequently used in the e-commerce sector. The customer list is to be implemented as a simple HTML table that is created using TypeScript. When working with AI, type definitions are always useful to specify the structure for which the AI should generate data. You can get help for generating such types by specifying the rough structure and letting AI take care of the details.

> **Prompt:** Please create a TypeScript alias `Customer` with the following fields: id, first name, last name, email, registration date, status, and last order date.

Sure! Here is how you can create a TypeScript type alias named `Customer` with the specified fields:

```
type Customer = {
  id: string;
  firstName: string;
  lastName: string;
  email: string;
  registrationDate: Date;
  // Assuming status is one of these values
  status: 'active' | 'inactive' | 'suspended';
  // Assuming lastOrderDate can be null if there's no order
  lastOrderDate: Date | null;
};
```

Here's a brief explanation of each field:

[explanation of the fields]

The generated code follows the instructions of the prompt exactly. For fields that aren't unique, such as `status` and `lastOrderDate`, a comment with the corresponding assumption has been inserted. `status` is a union type with three values, and either a date value or `null` can be used for `lastOrderDate`. You can restrict the prompt even further and specify certain types, but in this case, you lose the advantages of the AI tool, as you're about to implement the structure yourself. An alternative to such a prompt can be the auto-completion of tools such as GitHub Copilot. When you start implementing the customer type, Copilot suggests fields and data types that you can adopt.

The implementation consists of two parts: an HTML file, which is responsible for the display, and a TypeScript file, which builds up the data dynamically. The code of the HTML file could read as follows:

```
<!DOCTYPE html>
<html lang="en">
  <head>
    <meta charset="UTF-8" />
    <title>Document</title>
    <script type="module" src="list.js"></script>
  </head>
  <body>
    <table>
      <thead>
        <tr>
          <th>Id</th>
          <th>First name</th>
          <th>Last name</th>
          <th>Email</th>
          <th>Registration date</th>
          <th>status</th>
          <th>Last order date</th>
        </tr>
      </thead>
      <tbody></tbody>
    </table>
  </body>
</html>
```

The HTML structure consists of the general frame of the document, the header section, which loads the JavaScript file to build the table, and the body, which contains the basic structure of the table and an empty `<tbody>` element into which the data gets inserted.

```
import type { Customer } from './customer';

const customers: Customer[] = [];

document.addEventListener('DOMContentLoaded', () => {
  drawTable(customers);
});

function drawTable(customers: Customer[]) {
  const customerTable = customers.map((customer) => {
    return `
      <tr>
```

```
        <td>${customer.id}</td>
        <td>${customer.firstName}</td>
        <td>${customer.lastName}</td>
        <td>${customer.email}</td>
        <td>
          ${customer.registrationDate.toISOString()
             .split('T')[0]}
        </td>
        <td>${customer.status}</td>
        <td>
          ${customer.lastOrderDate.toISOString().split('T')[0]}
        </td>
      </tr>
    `;
  });

  document.querySelector('tbody')!.innerHTML =
    customerTable.join('');
}
```

What is still missing at this point is the test data. The browser typically loads this data asynchronously from the backend. For the sake of simplicity, however, we assume that the data is already available in the form of the customers constants. Here, you can start by inventing data yourself, but you can also ask the large language model (LLM) of your choice to generate test data for you.

Prompt: Please create an array with 10 elements of realistic test data. The data should represent American customers. Ensure that the lastOrderDate is later than the registrationDate. Use the following TypeScript type:

```
type Customer = {
  id: string;
  firstName: string;
  lastName: string;
  email: string;
  registrationDate: Date;
  status: 'active' | 'inactive' | 'suspended';
  lastOrderDate: Date | null;
};
```

Sure! Here is an array with 10 elements of realistic test data for American customers:

```
const customers: Customer[] = [
  {
    id: '1',
```

```
    firstName: 'Anna',
    lastName: 'Mueller',
    email: 'anna.mueller@example.de',
    registrationDate: new Date('2022-01-15'),
    status: 'active',
    lastOrderDate: new Date('2023-04-18'),
},
{
    id: '2',
    firstName: 'Luca',
    lastName: 'Rossi',
    email: 'luca.rossi@example.it',
    registrationDate: new Date('2021-11-20'),
    status: 'active',
    lastOrderDate: new Date('2023-02-25'),
},
...
```

You can use the response from ChatGPT directly in your application. The generated data corresponds to fairly realistic data; for example, the first and last name have been included correctly in the email address. The date values are also correct and follow the rule that the date of the last order should be after the date of registration.

This example shows how you can take some of the work out of generating test data. However, a little caution is also required here. The more freedom you give the AI tool, the more it can interpret. It was simply a matter of luck that the date of registration was before the date of the last order. The AI tool could also have reversed the date values here without any further specifications. This isn't a major problem for the list, but it can quickly become an issue for further work with the data, especially when it comes to validation routines.

In a case like this, you should also consider the number of data records that are generated. With a low number like 10, ChatGPT works reliably and error free. The more data records you request, the greater the risk that duplicates will creep in (which you can exclude via your prompt), or that the tool will refuse to work after a certain number of data records and either abort in the middle of the data structure or output something like "40 more data records may still follow here." In this case, you can have the data records generated in smaller groups and merge them yourself.

5.2 AI-Supported Test Automation

If you work on a new feature for your application, there will come a time when you need to check the functionality. A manual check is time consuming and prone to errors, especially if you consider that you also have to test at least parts of the rest of the application, as errors may have crept in at other points as a result of the integration. The automation potential for tasks such as checking the features of an application is huge. This is precisely why the discipline of test automation exists. You can test your application at various levels based on the *test pyramid*, which classifies the different types of tests at different levels:

- **Unit tests**
  The basis of the pyramid is formed by the unit tests. They check the smallest testable units of an application in isolation. These units are usually functions or methods. Unit tests are processed quickly, are easy to write, and are usually available in large numbers.

- **Integration tests**
  These tests form the middle level of the test pyramid. They test individual or several modules or components. Integration tests are more complex and slower than unit tests, but they also provide a higher degree of security as they test the interaction of multiple components in your application. Typically, your application has significantly fewer integration tests than unit tests.

- **End-to-end tests**
  End-to-end tests (sometimes referred to as E2E tests) form the top of the pyramid. They test features across the entire application, from the frontend to data management. These tests are the most complex to create and the slowest to run, but they provide the highest level of security. Compared to the other two levels, end-to-end tests typically take up less room.

AI tools can support you at all levels of the test pyramid and relieve you of some of the work involved in generating tests. In the remaining sections of this chapter, you'll learn about the different levels of the test pyramid and their characteristics using TypeScript as an example. However, tests work similarly in other programming languages and environments, so you can transfer your knowledge.

5.2.1   Creating AI-Supported Unit Tests

There are multiple testing frameworks for unit tests in JavaScript and TypeScript. One of the most popular is Vitest, a testing framework that comes from the Vue.js ecosystem, but is framework- and even platform-independent. That is why you can use the testing framework both on the client side in the browser and on the server side using Node.js. A major advantage of Vitest is that it doesn't require any configuration at all in simple cases. It automatically finds the tests based on their file name and executes them.

AI-Supported Unit Tests

The first step in the AI-supported generation of a unit test is a simple TypeScript function named add, which adds two integers and returns the result. The code for this function looks as follows:

```typescript
export default function add(a: number, b: number): number {
  if (!Number.isSafeInteger(a) || !Number.isSafeInteger(b)) {
    throw new Error("Can only add integers");
  }
  return a + b;
}
```

You can have ChatGPT generate a test for this function:

Prompt: Please write a Vitest unit test for the following function:

[code of the function]

Create only the test; the environment is already set up, and everything is installed.

```typescript
import { describe, it, expect } from 'vitest';
import add from './add'; // Adjust path to function add

describe('add', () => {
  it('should add two integers correctly', () => {
    expect(add(2, 3)).toBe(5);
    expect(add(-1, 1)).toBe(0);
    expect(add(0, 0)).toBe(0);
  });
  it(
    'should throw an error if either argument is not a safe ' +
    'integer', () => {
    expect(() => add(2.5, 3))
      .toThrow('Can only add integers');
    expect(() => add(2, 3.5))
      .toThrow('Can only add integers');
    expect(() => add(Number.MAX_SAFE_INTEGER + 1, 3))
      .toThrow('Can only add integers');
    expect(() => add(2, Number.MAX_SAFE_INTEGER + 1))
      .toThrow('Can only add integers');
  });

  it(
    'should throw an error if any argument is not a number',
    () => {
```

```
    expect(() => add('2', 3))
      .toThrow('Can only add integers');
    expect(() => add(2, '3'))
      .toThrow('Can only add integers');
    expect(() => add(null, 3))
      .toThrow('Can only add integers');
    expect(() => add(2, undefined))
      .toThrow('Can only add integers');
  });
});
```

[explanation of the scenarios]

If you specify the entire source code in the prompt, as in this example, the AI tool can cover all aspects of the functionality. The signature of the function contributes to this because here you specify exactly which data types you expect, and which type you return. The error handling within the function indicates additional limit and exception cases that must also be covered by the tests. AI tools such as ChatGPT like to be a little more detailed in their answers to such tasks and not only generate the requested test but also provide tips on setup and execution. You can prevent this by specifying that you're only interested in the test and not in the environment or the installation.

The result produced by ChatGPT is impressive. Not only does the generated code contain a simple test that adds the values 2 and 3, for example, and checks whether the correct result is returned, it contains a whole series of tests. What stands out positively here is that the tests are grouped using describe, and the group has been given the name of the function as a label. There are three tests within this group:

- **Successful case**
  The test expects the function to be able to add two integers. The first test ensures that two positive integers can be added, the second one adds a negative number, and the third test ensures that 0 and 0 can also be added.

- **Error case 1**
  The predefined implementation of the add function checks that both numbers are *safe integers*, that is, integers that lie in a range in which TypeScript or the JavaScript engine can calculate them correctly. The second test checks that calculating with floating point numbers and numbers that are too large (Number.MAX_SAFE_INTEGER + 1) isn't possible and leads to an exception.

- **Error case 2**
  The third test goes even further and ensures that calling the add function with invalid data types such as strings, null, or undefined leads to an error.

If successful, ChatGPT uses the toBe matcher to check whether the result matches the expected value. The choice of this check function fits well, as it checks internally for strict equality and doesn't allow type conversion. This function only causes difficulties if reference types such as objects are involved, which isn't the case here. The error cases correctly cover the tests using the toThrow matcher, which you can use to check whether a function throws an error.

Let's now move on to the weaknesses of the generated tests. A major problem is that the TypeScript compiler reports some errors in the tests when processing the code. The third test, which calls the function with different types, isn't possible in TypeScript because the types aren't compatible with the signature of the function when it's called. The prompt doesn't mention anything about TypeScript and type-safe code, but this should have been obvious when the function was analyzed.

You can address this problem by requiring ChatGPT to produce valid TypeScript code. To do this, you can either adapt the original prompt or specify the request as shown here:

Prompt: Make sure the tests are written in valid TypeScript. Keep the test if any argument isn't a number, but fix it.

Sure, here is the corrected test suite in valid TypeScript:

```
(existing code ...)
  it(
    'should throw an error if any argument is not a number',
    () => {
    expect(() => add('2' as any, 3))
      .toThrow('Can only add integers');
    expect(() => add(2, '3' as any))
      .toThrow('Can only add integers');
    expect(() => add(null as any, 3))
      .toThrow('Can only add integers');
    expect(() => add(2, undefined as any))
      .toThrow('Can only add integers');
  });
});
```

In this version: each invalid argument is explicitly cast to any to ensure the test case runs as expected while maintaining TypeScript type checks for valid arguments.

To be on the safe side, you should specify in the prompt that the third test should be retained in any case. This is unnecessary in a pure TypeScript environment, as you can't execute the function with an incorrect type. However, because TypeScript is only used at compile time and not at runtime, the wrong data types may still be used depending on the environment. The original two tests remain unchanged, as they were already written in valid TypeScript. To avoid the errors in the third test, ChatGPT uses Type-Script's type conversion and turns an invalid string data type into a number. Although this adaptation isn't elegant, it satisfies the TypeScript compiler.

Another point of criticism is that the tests don't just have one assertion, but always more than one. Although the tests always fit together thematically, there is often a requirement that each test should only have one assertion. The reason for this is that a test should only fail for one reason in order to facilitate troubleshooting. In this case, it's up to you to decide whether you would prefer to have fewer, thematically grouped tests, or many, finely granular tests. You can also solve this issue by extending the prompt and asking ChatGPT to create a separate test for each expectation:

Prompt: Please put every expectation in its own test. And keep the grouping by putting the tests in their own sub-describe blocks.

Sure, here is the revised test suite with each expectation in its own test and grouped using sub-describe blocks:

```
import { describe, it, expect } from 'vitest';
import add from './add'; // Adjust path to function add

describe('add', () => {
  describe('valid inputs', () => {
    it('should add 2 and 3 correctly', () => {
      expect(add(2, 3)).toBe(5);
    });

    ... more tests ...
  });

  describe('invalid safe integer inputs', () => {
    ... more tests ...
  });

  describe('invalid non-number inputs', () => {
    ... more tests...
  });
});
```

[explanation of the code]

If you only prompt the first sentence, ChatGPT turns the original 3 tests into a total of 11 tests. The second sentence of the prompt ensures that the grouping gets maintained and that you don't lose the overview so quickly.

In general, you'll have to deal with the topic of testing, although tools such as ChatGPT, Copilot, and the like can do a lot of work for you. You should familiarize yourself with the structure and organization of tests and develop a basic knowledge of test patterns and the architecture of tests. The more precisely you formulate your prompts, the more likely the result will meet your expectations, and the code of the tests will remain readable and maintainable in the long term.

However, the possibilities of AI don't end with the formulation of tests; the tools can also support you in test-driven development.

5.3   Test-Driven Development with AI

Test-driven development (TDD) is a cyclical process that is often described as a TDD cycle. It consists of three phases: red, green, and refactor. The first step is a failed test, as it occurs before the implementation. The red color stands for the failure of the test. You then implement the program logic that ensures the test can run successfully. This turns the test into green. In the final step, you have the opportunity to optimize your code. You carry out refactoring in which the interface to the outside world remains the same and nothing changes in terms of functionality. Chapter 4 looks at how AI can help you with this and what strategies are available.

In TDD, you don't start directly with the development. Instead, first think about the requirements that your code unit, usually a function, should fulfill, and formulate these in the form of one or more tests. Then, you implement the function. You proceed in small steps, so-called baby steps, and implement the function step by step or, rather, test by test.

In this chapter, we use two different AI tools for TDD, but in real life, you'll rarely do this. This example is rather intended to show how you can get help with the various parts of the development process.

As a concrete example of TDD, we're implementing a function for checking credit card numbers. Such a number is made up of different parts: the first six digits identify the issuing financial institution, while the subsequent digits form the cardholder's account number. The last digits, which are relevant for our example, form the check digit. This is calculated using the Luhn algorithm.

Because we want the AI tool to do the work for us and not keep us unnecessarily busy, we have two tools working together here: ChatGPT writes the test, and GitHub Copilot takes care of the implementation.

5.3.1   Implementing the First Test

Tools like ChatGPT are very useful and usually try to give a comprehensive answer. However, this is more of a hindrance for our current task. To prevent the tool from showing you how to set up your test environment, deliver all the tests and implementation, and run the tests, you need to make the prompt very clear:

> **Prompt:** Assume the environment is already set up and everything is installed. Create a test in `vitest` using TypeScript for a function called `validateCreditCardNumber` to validate a given credit card number. Use TDD and baby steps as a strategy. Only write the first test; no info on how to run the test is needed.

> Sure, let's start with the first step in Test-Driven Development (TDD): writing a test for a nonexisting function `validateCreditCardNumber`.
>
> Step 1: Write the First Test. We'll write a test that checks if the `validateCreditCardNumber` function exists and is callable. Let's create a test file `validateCreditCardNumber.test.ts`:
>
> ```
> import { describe, it, expect } from 'vitest';
> import {
>   validateCreditCardNumber
> } from './validateCreditCardNumber';
>
> describe('validateCreditCardNumber', () => {
>   it('should be a function', () => {
>     expect(typeof validateCreditCardNumber).toBe('function');
>   });
> });
> ```
>
> [Explanation of the code]

With this prompt, ChatGPT keeps the response concise and focused. The generated code imports the functions required by Vitest as well as the implementation of the `validateCreditCardNumber` function. The first test only checks whether there really is a function behind this variable. If you save the code in a file named `validateCreditCardNumber.test.ts`, as suggested by ChatGPT, and then run the test via `npx vitest`, you'll receive an error message indicating that the `validateCreditCardNumber` file doesn't exist. So, in the next step, you create the `validateCreditCardNumber.ts` file.

You can then instruct GitHub Copilot to generate the implementation for you. You can achieve this with the following prompt, for example:

> **Prompt:** Make the following test pass:
> [code of the test]

Sure! Here's the code block you can use to make the test pass:

```
export function validateCreditCardNumber(
  cardNumber: string
): boolean {
  // Your implementation here
  return true;
}
```

Running the test with this implementation returns a green result, so it's time for the next test.

GitHub Copilot has enhanced its capabilities to reference other files within a project. In Visual Studio and Visual Studio Code, you can reference specific files in GitHub Copilot Chat by using the # symbol followed by the file name. For example, to reference a file named `validateCreditCardNumber.ts`, you would use `#validateCreditCardNumber.ts`.

Additionally, GitHub has introduced Copilot Workspace, a Copilot-native development environment designed to assist developers with tasks ranging from brainstorming to coding and testing. Copilot Workspace enables AI-assisted development within a repository, allowing for programming tasks embedded in a larger context.

5.3.2   The Second Test

If the first test is still in the context window of the ChatGPT session, you can start from the previous prompt and ask ChatGPT to implement the next test:

Prompt: Create the next test.

Step 2: Write the Next Test. We will now write a test to check if the `validateCreditCardNumber` function correctly validates a valid credit card number. Here's an updated version of `validateCreditCardNumber.test.ts` with the next test:

```
import { describe, it, expect } from 'vitest';
import {
  validateCreditCardNumber
} from './validateCreditCardNumber';

describe('validateCreditCardNumber', () => {
  (existing code)

  it('should return true for a valid credit card number', () => {
    // Example of a valid Visa card number
    const validCreditCardNumber = '4111111111111111';
    expect(validateCreditCardNumber(validCreditCardNumber))
```

```
      .toBe(true);
  });
});
```
[explanation of the code]

The second test already includes significantly more framework conditions. It transfers a credit card number as a character string and expects the true value as the return value. If you run the tests, you'll find that both are already successfully executed without changing the source code. The reason for this is that the function always returns the true value. So, for now, no further changes to the code are necessary, but the next test is due immediately.

5.3.3   The Third Test

Now, it's ChatGPT's turn again with the next test. Due to the existing context, this prompt is also kept very simple:

Prompt: Create the next test.

Sure, let's proceed by writing the next test for the validateCreditCardNumber function. This test will check if the function correctly identifies an invalid credit card number. Here's an updated version of validateCreditCardNumber.test.ts with the next test:

```
import { describe, it, expect } from 'vitest';
import {
  validateCreditCardNumber
} from './validateCreditCardNumber';

describe('validateCreditCardNumber', () => {
  (existing code)

  it(
    'should return false for an invalid credit card number',
    () => {
    // Example of an invalid credit card number
    const invalidCreditCardNumber = '1234567890123456';
    expect(validateCreditCardNumber(invalidCreditCardNumber))
      .toBe(false);
  });
});
```
[explanation of the code]

This third test is a classic negative test. It allows you to test whether the function also handles incorrect entries correctly. This also forces you to adapt the implementation of the function so that the algorithm is correct. In the prompt, you pass the three tests to GitHub Copilot and instruct it to change the code accordingly.

> **Prompt:** Change the implementation, so that the following three tests pass:
>
> [code of the test]

```
// Implementation of validateCreditCardNumber function
export function validateCreditCardNumber(
  cardNumber: string
): boolean {
  // Remove any non-digit characters from the card number
  const cleanedCardNumber = cardNumber.replace(/\D/g, "");

  // Check if the cleaned card number is empty or not
  if (cleanedCardNumber.length === 0) {
    return false;
  }

  // Use Luhn algorithm to validate the card number
  let sum = 0;
  let double = false;

  for (let i = cleanedCardNumber.length - 1; i >= 0; i--) {
    let digit = parseInt(cleanedCardNumber.charAt(i), 10);

    if (double) {
      digit *= 2;
      if (digit > 9) {
        digit -= 9;
      }
    }

    sum += digit;
    double = !double;
  }

  return sum % 10 === 0;
}
```

Copilot completed the task with flying colors and wrote the code in such a way that all three tests could be passed successfully. The individual steps in the function are provided with descriptive comments so that you can easily follow the code.

The step from the empty function to the final implementation doesn't correspond to the idea of TDD. Further intermediate steps would have been advisable here. In this case, you should have been much stricter in the design of the tests and also in the prompting for Copilot and guided the AI tool better. Again, the following also applies here: the more precisely you formulate your prompts, the better you can control the result.

5.3.4   Generating Test Data

If you're still skeptical as to whether the implementation really works, you can have another test generated that checks a whole range of credit card numbers. For such a mass test, you can use the it.each method to test a range of values and still keep your code compact. The corresponding prompt could read as follows:

> **Prompt:** Create 10 valid and 10 invalid credit card numbers, and use it.each to verify if the validateCreditCardNumber function works correctly.

Sure, let's enhance our tests using it.each to validate multiple credit card numbers at once. We'll create 10 valid and 10 invalid credit card numbers and use it.each to verify if the validateCreditCardNumber function works correctly.

Here's the updated validateCreditCardNumber.test.ts:

```
import { describe, it, expect } from 'vitest';
import {
  validateCreditCardNumber
} from './validateCreditCardNumber';

const validCreditCardNumbers =
  [... validCreditCardNumbers ... ];

const invalidCreditCardNumbers =
  [ ... invalidCreditCardNumbers ... ];

describe('validateCreditCardNumber', () => {
  it('should be a function', () => {
    expect(typeof validateCreditCardNumber).toBe('function');
  });

  it.each(validCreditCardNumbers)(
    'should return true for a valid credit card number: %s',
```

```
   (cardNumber) => {
   expect(validateCreditCardNumber(cardNumber)).toBe(true);
 });

 it.each(invalidCreditCardNumbers)(
   'should return false for an invalid credit card
    number: %s',
   (cardNumber) => {
   expect(validateCreditCardNumber(cardNumber)).toBe(false);
 });
});
```

[explanation of the code]

Due to the existing context, ChatGPT has understood this prompt as an improvement, that is, refactoring of the existing tests. The implementation of validateCreditCardNumber remains untouched, but the code of the tests changes in this case. Generally, refactoring of the tests is also permitted. In this example, you define two arrays, one with valid credit card numbers and one with invalid ones.

The first test as to whether validateCreditCardNumber is a function remains unchanged. For the valid and invalid credit card numbers, you use the it.each function to generate a separate test for each value. The it.each function uses placeholders in the description so that the respective value gets inserted in the output.

In the next section, you'll learn how to deal with external dependencies in your tests and how AI tools can help you with this.

5.4   Using Dependencies

Modern test frameworks provide the option of defining spies, stubs, and mocks, which are all also called test doubles. You can think of this as a function whose calls you can check (using the *spy*), whose behavior you can predefine (using the *stub*), or whose interaction rules you can define and then check whether they have been adhered to at the end of a test (using the *mock*). Spies and stubs in particular are very useful when it comes to formulating unit tests. If you use callback functions, such as event handlers, you can't avoid using spy functions.

In the following sections, we'll focus on the use of stubs to write unit tests for functions that have additional dependencies. In a unit test, you focus on a specific unit of your source code, usually a function or method. As soon as this works together with other parts of your application, the situation becomes more complex. For one thing, the test can no longer fail because of just one function, and for another, you potentially have to cover multiple flow paths, which means that there is a larger number of tests to be performed.

To make the whole thing more concrete, we'll assume a TypeScript class named Order-Service in the following example. This has a placeOrder method that enables you to save an order. When saving, your application enriches the order with additional information and uses another class—the OrderRepository class—to persist the order. The greatly simplified source code of these two classes can be found in the following code example:

```typescript
import { randomUUID } from "node:crypto";

type Item = {
  id: string;
  name: string;
  price: number;
  quantity: number;
};
type Order = {
  id: string;
  items: Item[];
  total: number;
  status: "new" | "pending" | "completed" | "cancelled";
  createdAt?: Date;
};

export class OrderRepository {
  async save(order: Order): Promise<Order> {
    return {
      ...order,
      id: randomUUID(),
      createdAt: new Date(),
    };
  }
}

export class OrderService {
  constructor(private orderRepository: OrderRepository) {}

  placeOrder(order: Order) {
    const total = order.items.reduce((acc, item) => {
      return acc + item.price * item.quantity;
    }, 0);

    const orderRecord: Order = {
```

```
    ...order,
    total,
    status: "pending",
  };

  return this.orderRepository.save(orderRecord);
  }
}
```

When it gets instantiated, OrderService receives an instance of the OrderRepository class. If you call the placeOrder method with an Order object, the method enriches this object with the total amount of the ordered parts and sets the status to pending. The method then calls the save method of the OrderRepository object. This method in turn assigns a unique ID to the order and sets the creation date.

The AI tool of our choice—in this case, ChatGPT—should generate a test for the placeOrder method and ensure that the save method of the OrderRepository isn't called, but rather is replaced by a stub method.

Prompt: Write a Vitest test for the placeOrder method, and replace the save method of the OrderRepository with a stub.

[code of the example]

Here's how you can write a test for the placeOrder method in your OrderService class using Vitest. In this test, we'll replace the save method of the OrderRepository with a stub.

First, ensure that you have Vitest installed and set up in your project. Then, create a test file, for example, OrderService.test.ts, and add the following code:

```
import { describe, it, expect, vi } from 'vitest';
import {
  OrderService, OrderRepository, Order, Item
} from './path-to-your-order-service';

describe('OrderService', () => {
  it('should place an order and calculate the total',
    async () => {
    // Create a stub for the OrderRepository save method
    const saveStub = vi.fn().mockResolvedValue({
      id: 'stubbed-id',
      items: [],
      total: 100,
      status: 'pending',
      createdAt: new Date(),
```

```
  });

  // Create a mock OrderRepository
  const orderRepository = new OrderRepository();
  orderRepository.save = saveStub;
  // Instantiate the OrderService with the mock
  // OrderRepository
  const orderService = new OrderService(orderRepository);

  // Define a sample order
  const order: Order = {
    id: '',
    items: [
      { id: '1', name: 'item1', price: 10, quantity: 5 },
      { id: '2', name: 'item2', price: 20, quantity: 2 },
    ],
    total: 0,
    status: 'new',
  };

  // Place the order
  const result = await orderService.placeOrder(order);

  // Assertions
  expect(saveStub).toHaveBeenCalledOnce();
  expect(saveStub).toHaveBeenCalledWith(
    expect.objectContaining({
      total: 90,
      status: 'pending',
    }));
  expect(result).toEqual(expect.objectContaining({
    id: 'stubbed-id',
    total: 90,
    status: 'pending',
  }));
  });
});
[explanation of the test]
```

The test first prepares the environment as required by creating a stub for the save method. It then creates a new OrderRepository, inserts the stub there, and generates a new instance of OrderService. The most important part of the test is the call of the `placeOrder` method. To do this, the test defines an `order` object that represents the order and passes it when the method is called. The test concludes with the following checks:

- Has the `saveStub` been called exactly once?
- Was the `saveStub` called with the correct object?
- Is the return value of the `placeOrder` method correct?

The test looks good so far, and you can integrate and execute it directly in your application. However, you'll then see a significant problem because the test fails when checking the return value of the `placeOrder` method. The reason for this is that the stub function returns the value 100, for example, and not 90 as required. The `items` and `createdAt` properties are also missing.

You can solve this problem by copying the error message to the next prompt. In this case, ChatGPT revises the answer and provides you with a functioning test.

Prompt: I got this error, please fix the test:

```
AssertionError: expected { id: 'stubbed-id',
                            items: [], ...(3) }
                to deeply equal ObjectContaining{...}
- Expected
+ Received

- ObjectContaining {
+ Object {
+    "createdAt": 2024-07-21T14:18:23.915Z,
     "id": "stubbed-id",
+    "items": Array [],
     "status": "pending",
-    "total": 90,
+    "total": 100,
  }
 > src/order.service.test.ts:49:20
    47|        })
    48|      );
    49|      expect(result).toEqual(
      |                     ^
    50|        expect.objectContaining({
    51|          id: "stubbed-id"
```

The error indicates a mismatch between the expected and received objects in the assertion. Specifically, the total value and `createdAt` field are different. Let's update the test to correctly handle these discrepancies. Here is the updated test code:

```
import { describe, it, expect, vi } from 'vitest';
import {
  OrderService, OrderRepository, Order, Item
} from './path-to-your-order-service';
describe('OrderService', () => {
  it('should place an order and calculate the total',
    async () => {
    // Create a stub for the OrderRepository save method
    const saveStub = vi.fn()
      .mockImplementation(async (order: Order) => ({
        ...order,
        id: 'stubbed-id',
        createdAt: new Date(),
      }));

    // Create a mock OrderRepository
    const orderRepository = new OrderRepository();
    orderRepository.save = saveStub;

    // Instantiate the OrderService with the mock OrderRepository
    const orderService = new OrderService(orderRepository);

    // Define a sample order
    const order: Order = {
      id: '',
      items: [
        { id: '1', name: 'item1', price: 10, quantity: 5 },
        { id: '2', name: 'item2', price: 20, quantity: 2 },
      ],
      total: 0,
      status: 'new',
    };

    // Place the order
    const result = await orderService.placeOrder(order);

    // Assertions
    expect(saveStub).toHaveBeenCalledOnce();
    expect(saveStub).toHaveBeenCalledWith(expect
      .objectContaining({
        total: 90,
        status: 'pending',
      }));
```

```
    expect(result.id).toBe('stubbed-id');
    expect(result.total).toBe(90);
    expect(result.status).toBe('pending');
    expect(result.items).toEqual(order.items);
    expect(result.createdAt).toBeInstanceOf(Date);
  });
});
```

In this updated test:

1. The `saveStub` now uses `mockImplementation` to return the `order` object with the `id` and `createdAt` fields added, simulating the actual save operation more closely.

2. The assertions are broken down to individually check each field, making the test clearer and avoiding deep equality checks that might fail due to differences in other fields (like `createdAt`).

ChatGPT has tackled precisely the two critical points when adapting the source code. The stub implementation is now based on the transferred object and only adds two properties. When checking the return value, the test now proceeds property by property so that no more errors can occur here either. If you integrate this test into your environment, it will run without an error message.

5.5   Optimizing Tests

AI tools can help you not only implement tests but also improve them. You've already seen an example of this when it came to implementing a whole series of tests for an array of values. But the possibilities don't end there. In the following sections, we take a look at two additional use cases: the improvement of existing tests and the implementation of missing tests.

5.5.1   Optimizing Test Code

When implementing tests, it occasionally happens that the tests correctly check the desired functionality and thus fulfill their purpose, but aren't optimally implemented. In general, the same rules apply to tests as to all other code: They should meet the quality criteria of the project. AI tools can make your work easier here by improving the code of your tests. However, this doesn't mean that you no longer need to pay attention to the quality of your source code. Rather, in this case, AI is an indicator of where there is still room for improvement.

The `UserService` TypeScript class serves as an example here. It manages an array of user objects and can add, read, and delete them. The implementation reads as follows:

```
export type User = {
  id: number;
  name: string;
};

export class UserService {
  private users: User[];

  constructor() {
    this.users = [];
  }
  addUser(user: User): User {
    this.users.push(user);
    return user;
  }
  getUserById(id: number): User | undefined {
    return this.users.find(user => user.id === id);
  }
  deleteUserById(id: number): User | null {
    const index = this.users.findIndex(user => user.id === id);
    if (index !== -1) {
      return this.users.splice(index, 1)[0] || null;
    }
    return null;
  }
}
```

For this class, you can write a series of tests like the following:

```
import { expect, it } from 'vitest';
import { UserService } from './UserService';
import type { User } from './UserService';

it('addUser should add a user to the list', () => {
  const userService = new UserService();
  const user: User = { id: 1, name: 'Alice' };
  userService.addUser(user);
  expect(userService['users']).toContain(user);
});

it('getUserById should return the correct user', () => {
  const userService = new UserService();
  const user1: User = { id: 1, name: 'Alice' };
  const user2: User = { id: 2, name: 'Bob' };
```

```
  userService.addUser(user1);
  userService.addUser(user2);
  expect(userService.getUserById(1)).toBe(user1);
  expect(userService.getUserById(2)).toBe(user2);
});

it('getUserById should return undefined for non-existing user', () => {
  const userService = new UserService();
  expect(userService.getUserById(99)).toBeUndefined();
});

it('deleteUserById should delete the correct user', () => {
  const userService = new UserService();
  const user1: User = { id: 1, name: 'Alice' };
  const user2: User = { id: 2, name: 'Bob' };
  userService.addUser(user1);
  userService.addUser(user2);
  userService.deleteUserById(1);
  expect(userService['users']).not.toContain(user1);
  expect(userService['users']).toContain(user2);
});

it('deleteUserById should return null for non-existing user', () => {
  const userService = new UserService();
  expect(userService.deleteUserById(99)).toBeNull();
});
```

These tests cover all three cases, that is, adding, reading, and deleting. Nevertheless, the tests provide potential for improvement:

- **Duplicates with objects**
  There are several instances of the UserService class and the user objects.

- **No use of setup and teardown routines**
  The environment can be initialized before the tests and cleaned up after the tests.

- **No clean structure**
  The test code is not organized, making it difficult to understand, maintain, and extend.

You can easily solve the problem with an adequate prompt. Here, the same principle applies that the more precisely you specify what you expect, the better the result will be. For example, you can indicate that you expect a clean structure, that duplicates in the code should be eliminated as much as possible, and that the best practices of clean code should generally be followed.

Prompt: Could you please optimize the following Vitest test that is written in Type-Script? Introduce a clean structure for the tests, eliminate duplicates, and use the clean code principles.

[code of the tests]

To optimize the given tests, we can introduce a common setup function, use a consistent naming convention, and remove duplicate code. Additionally, we can make sure that each test is focused on a single behavior. Here's the refactored code:

```typescript
import { expect, it, describe, beforeEach } from 'vitest';
import { UserService } from './userService';
import type { User } from './userService';

describe('UserService', () => {
  let userService: UserService;
  let user1: User;
  let user2: User;

  beforeEach(() => {
    userService = new UserService();
    user1 = { id: 1, name: 'Alice' };
    user2 = { id: 2, name: 'Bob' };
  });

  it('should add a user to the list', () => {
    userService.addUser(user1);
    expect(userService['users']).toContain(user1);
  });

  it('should return the correct user by ID', () => {
    userService.addUser(user1);
    userService.addUser(user2);
    expect(userService.getUserById(1)).toBe(user1);
    expect(userService.getUserById(2)).toBe(user2);
  });

  it('should return undefined for non-existing user
      by ID', () => {
    expect(userService.getUserById(99)).toBeUndefined();
  });

  it('should delete the correct user by ID', () => {
    userService.addUser(user1);
    userService.addUser(user2);
```

```
    userService.deleteUserById(1);
    expect(userService['users']).not.toContain(user1);
    expect(userService['users']).toContain(user2);
  });

  it('should return null when deleting a non-existing user
      by ID', () => {
    expect(userService.deleteUserById(99)).toBeNull();
  });
});
```

[explanation of the code]

The revised tests are of significantly better quality. They are contained in a `describe` block that defines a context for the tests. The `userService`, `user1`, and `user2` variables are reused by the tests, and the setup routine in the `beforeEach` function ensures that the service and the user objects are instantiated correctly. This means that each test has a clean environment, and dependencies between the tests are eliminated.

Nevertheless, there is one small point of criticism. In the tests for `getUserById` and `deleteUserById`, the `addUser` method is used to write data to the service. If this method is incorrect, the other tests will also fail, whereas a test should generally only fail for one reason. A more elegant option here would be to write the information directly to the service or to transfer the data during initialization.

5.5.2   Writing Missing Tests

When developing your application, you should try to validate the entire code using automated tests. However, in some cases, this isn't possible or is too time consuming. Then, you should at least check the critical points. One tool that is very useful here is the *coverage report*. During test execution, it measures which lines, statements, and branches in the source code are checked by the tests. In general, coverage of 80% is the target value. If your work is test-driven, this value can reach up to 100%.

Most modern testing frameworks support coverage reports out of the box or as extensions. With Vitest, the framework we're using in this chapter, you can choose between the V8 and the Istanbul reporter, with the former being the default. To generate a report, you need to install the extension using the `npm install -D @vitest/coverage-v8` command and then run your tests by entering the `vitest run --coverage` command. Vitest then generates the coverage report in a series of HTML files that you can view in your browser.

As a concrete example, we use the `UserService` class from the previous example again, but with a slightly reduced range of functions. The class now only has the `addUser` and getUserById methods. Of the two methods, only the addUser method is covered by a test. Figure 5.1 shows the coverage report for the file.

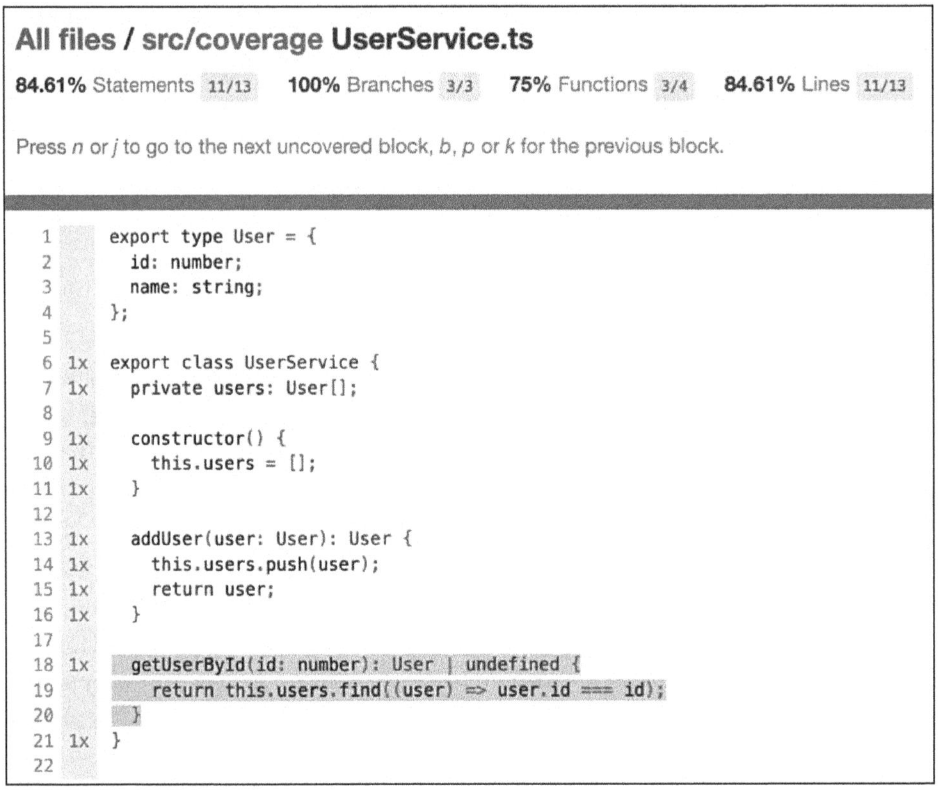

```
All files / src/coverage UserService.ts

84.61% Statements 11/13    100% Branches 3/3    75% Functions 3/4    84.61% Lines 11/13

Press n or j to go to the next uncovered block, b, p or k for the previous block.
```

```
 1       export type User = {
 2         id: number;
 3         name: string;
 4       };
 5
 6  1x   export class UserService {
 7  1x     private users: User[];
 8
 9  1x     constructor() {
10  1x       this.users = [];
11  1x     }
12
13  1x     addUser(user: User): User {
14  1x       this.users.push(user);
15  1x       return user;
16  1x     }
17
18  1x     getUserById(id: number): User | undefined {
19           return this.users.find((user) => user.id === id);
20         }
21  1x   }
22
```

Figure 5.1 Coverage Report for the UserService Class

You can now transfer the HTML code of the report to ChatGPT, for example, and have the missing tests generated.

> **Prompt:** Please generate the missing tests in Vitest using TypeScript for the following coverage report:
>
> [code of the coverage report]
>
> Here are the already existing tests, please extend them:
>
> [code of the existing tests]

Here is the extended test suite for the UserService class, covering the getUserById method, which is not currently tested.

```
import { expect, it, describe, beforeEach } from "vitest";
import { UserService } from "./UserService";
import type { User } from "./UserService";
```

```
describe("UserService", () => {
  let userService: UserService;
  let user1: User;
  let user2: User;

  beforeEach(() => {
    userService = new UserService();
    user1 = { id: 1, name: "Alice" };
    user2 = { id: 2, name: "Bob" };
  });

  it("should add a user to the list", () => {
    userService.addUser(user1);
    expect(userService["users"]).toContain(user1);
  });

  it("should return a user by ID", () => {
    userService.addUser(user1);
    const result = userService.getUserById(1);
    expect(result).toEqual(user1);
  });

  it("should return undefined if user ID does not exist", () => {
    userService.addUser(user1);
    const result = userService.getUserById(2);
    expect(result).toBeUndefined();
  });

  it("should handle multiple users correctly", () => {
    userService.addUser(user1);
    userService.addUser(user2);
    const result1 = userService.getUserById(1);
    const result2 = userService.getUserById(2);
    expect(result1).toEqual(user1);
    expect(result2).toEqual(user2);
  });
});
```

The AI tool has correctly recognized from the coverage report that the getUserById method isn't sufficiently tested and has generated a total of three additional tests for this method. The basis for the generated tests is the existing test file, into which the new tests are seamlessly embedded. The new tests don't just call the getUserById method, but they also check whether it's possible to read out a user, what happens if the requested object doesn't exist, and whether it's possible to read out several users.

In this case, the AI tool makes it much easier for us to generate the tests. If, as in this example, you pass additional context in the form of existing tests, the result will be even better, and you can transfer the generated code directly into your application without making any changes.

5.6   End-to-End Tests

Theoretically, testing frameworks such as Vitest enable you to cover all levels of the test pyramid. This class of frameworks is best suited for the formulation of unit and integration tests. There are separate frameworks for end-to-end tests, such as Selenium, Cypress, or Playwright, which have specialized in making the implementation of end-to-end tests as convenient as possible for you.

In the following sections, you'll learn how to use Playwright to generate an end-to-end test for a simple login form. The form has an input field for the username and one for the password. There is also a button for sending the form. The form is transmitted via a JavaScript function. Upon successful login, users are redirected to the /list path. The HTML code of the form reads as follows:

```html
<!DOCTYPE html>
<html lang="en">
  <head>
    <meta charset="UTF-8" />
    <meta
      name="viewport"
      content="width=device-width, initial-scale=1.0"
    />
    <title>Login Form</title>
  </head>
  <body>
    <div class="login-container">
      <h2>Login</h2>
      <form id="loginForm">
        <input
          type="text"
          id="username"
          name="username"
          placeholder="Username"
          required
        />
        <input
          type="password"
          id="password"
          name="password"
```

```
      placeholder="Password"
      required
    />
    <button type="submit">Login</button>
  </form>
 </div>
 </body>
</html>
```

We deliberately omit the styling and JavaScript here because the end-to-end test is a classic black box test that checks the functionality from the user's perspective.

5.6.1   Generating an End-to-End Test

The prerequisite for your end-to-end test to work is an installation of Playwright, which you can achieve via the `npx init playwright@latest` command. You can run the Playwright tests using the `npx playwright test` command. However, only sample tests are available immediately after installation. But you can change this by asking ChatGPT to generate a test for the login form for you.

For the test to work, the AI tool needs some information from you. You must provide the HTML structure and specify the address via which the form can be accessed. You must also specify which user name and password should be used and where you'll be redirected to after successful login. The corresponding prompt could look as follows:

Prompt: Create a Playwright test for the following HTML structure. Assume there is JavaScript that handles the form submission. The form can be found at *http://localhost:3000/login*. Use "admin" as username and "test" as password. After the successful login, the user will be redirected to *http://localhost:3000/list*.

[HTML form code]

Here is a Playwright test for the provided HTML structure. This test script will fill in the username and password fields with "admin" and "test" respectively, submit the form, and check if the user is redirected to http://localhost:3000/list.

First, make sure you have Playwright installed. If not, you can install it using:

`npm install @playwright/test`

Next, create a test file (e.g., login.test.js) and add the following code:

```
// @playwright/test module
import { test, expect } from "@playwright/test";

test("login test", async ({ page }) => {
```

175

```
// Go to the login page
await page.goto("http://localhost:3000/login");

// Fill in the username
await page.fill("#username", "admin");

// Fill in the password
await page.fill("#password", "test");

// Click the login button
await page.click('button[type="submit"]');

// Check if the the redirected URL is the list page
await page.waitForURL("http://localhost:3000/list");

// Check that the new URL is the list page
expect(page.url()).toBe("http://localhost:3000/list");
});
```
[explanation of the code]

Create a new file named login.test.js in the test directory of your application, and then run the test command as suggested in the response to the prompt. If the test runs successfully, it checks your login process. ChatGPT has also provided the individual lines with comments so you can easily follow the steps. You can find out even more about what the individual commands mean in the following explanation.

One limitation that you have to live with, at least with LLMs such as ChatGPT, is that you can't cover an entire workflow in one step with an end-to-end test. The problem here is that the AI tool can't dynamically run through a workflow that consists of multiple views. For this reason, you need to proceed step by step and then put the individual parts together to form an overall workflow.

5.6.2   Creating a Page Object

Your tests will be even clearer if you work with page objects. A *page object* provides you with speaking methods that enable you to interact with a page. The methods can have both read and write access to the page and, for example, return a specific value for checking or write values to a form field. This makes the test code even easier to read.

To use a page object in your end-to-end test, use the previous ChatGPT session, and specify the following as the next prompt:

Prompt: Could you include a page object in the test?

Certainly! Using the Page Object Model (POM) helps to organize and maintain your test code more efficiently. Below is an example that includes a page object for the login page and incorporates it into the test.

First, create a new file named `LoginPage.js` for the page object.

```
// LoginPage.js
export class LoginPage {
  /**
   * @param {import('@playwright/test').Page} page
   */
  constructor(page) {
    this.page = page;
    this.usernameInput = '#username';
    this.passwordInput = '#password';
    this.submitButton = 'button[type="submit"]';
  }
  async goto() {
    await this.page.goto('http://localhost:3000/login');
  }
  async login(username, password) {
    await this.page.fill(this.usernameInput, username);
    await this.page.fill(this.passwordInput, password);
    await this.page.click(this.submitButton);
  }
}
```

Next, update the test file to use this page object.

```
// login.test.js
import { test, expect } from '@playwright/test';
import { LoginPage } from './LoginPage';

test('login test', async ({ page }) => {
  const loginPage = new LoginPage(page);

  // Navigate to the login page
  await loginPage.goto();

  // Perform the login action
  await loginPage.login('admin', 'test');

  // Check if the the redirected URL is the list page
  await page.waitForURL("http://localhost:3000/list");
```

```
// Check that the new URL is the list page
expect(page.url()).toBe('http://localhost:3000/list');
});
```
[explanation of the code]

If it's purely a question of the number of lines, you've gained nothing by using the page object. The three lines that are lost by filling out the form are lost again by importing, instantiating the page object, and calling the methods. In a larger test, however, the code is usually much more compact and readable. You can also reuse the page objects between different end-to-end tests and gain from this. For example, if you swap out the login process to such a page object, you can use it in every end-to-end test in which you have to log in.

5.7   Conclusion

The big problem with test automation is that it takes place in the background and, unlike actual feature development, isn't visible in the product. For this reason, the automated testing of an application is often neglected. Using AI tools such as ChatGPT or GitHub Copilot, you can have a lot of work done for you and thus become significantly faster in the implementation of tests. The tools can also help you refactor tests to optimize them.

If you have no experience in formulating tests, AI will help you take the first steps. In addition to the generated code, you'll also receive explanations of the purpose of the individual parts of a test. If you've already implemented tests and have experience with the tools, architectures, and design patterns, you can incorporate these into your prompts, which will have a positive impact on the result. No matter how much experience you already have, you save a lot of development time compared to writing the tests manually.

Chapter 6
Documenting Software

Documentation is an indispensable but often unloved part of software development. The documentation of an application can be divided into the following different levels. It differs in its target audience and its proximity to the source code.

- **User manual**
  This type of documentation is intended for the end users and support staff of the software and is therefore also relatively far removed from the source code of the application. A user manual is used to explain the features of the application, describe how the application can be installed if necessary, what problems frequently occur, and how they can be solved.

- **Technical specification**
  The technical specification is aimed at developers, architects, and technical project managers. It has no direct reference to the code, but describes the architecture, components, and interfaces of the project. Data models, algorithms, and technologies used in the application are also included. The specification provides a good overview of the application and the interrelationships in the project.

- **API documentation**
  The application programming interface (API) documentation depends on the type of software you create. If you work on a library, this documentation usually describes the public functions developers use when integrating the library. In the case of a web service, the API documentation describes the interfaces your service makes available to the outside world. The API documentation is always intended for developers who want to connect to your software and is therefore directly related to a part of your source code.

- **Class and function documentation**
  This type of documentation is a direct component of your source code and is aimed at developers who work directly with the code base and extend and maintain it. In the case of classes, the comments describe the purpose and everything worth knowing about the class. Function documentation describes the purpose of a function, the parameters, and the return value, as well as relevant implementation details.

- **Inline comments**
  Inline comments have the most direct reference to the source code of all documentation variants; they are embedded in the code, either as separate lines or even as part of a program line. They are used to provide developers with direct information about the code.

In this chapter, we focus on the documentation types with a direct reference to the source code, that is, API documentation, class and function documentation, and inline comments. These types of documentation are mainly used to communicate information between members of the development team. Opinions differ widely as to which parts of the source code should be documented and how.

The different camps range from "every block in the code must be documented" to "my code is self-explanatory, so documentation is unnecessary." We want to choose a pragmatic middle way: all public interfaces that other developers have to deal with should be documented. In addition, complex code passages should be provided with explanatory comments.

6.1   Challenges with Documentation

Documentation, like automated testing, isn't necessarily a favorite discipline of developers. As with testing, there are also a number of challenges with documentation that you can overcome with the help of AI support. These include the following:

- **Time required**
  Good documentation takes time. In function documentation, for example, it's not enough to simply repeat the function name and list the names of the parameters. AI tools can do the work for you here by generating the documentation for you.

- **Continuous updating**
  There are few things worse than outdated documentation. Many developers rely on the documentation in the code. If that isn't up-to-date, avoidable errors can occur. For this reason, you must also check the documentation every time you change the code and adapt it if necessary. You can also outsource this task, at least in part, to the AI tool.

- **Consistency and accuracy**
  If different people work on the source code, inconsistencies can creep in. Everyone has their own way of documenting; if it differs too much from that of others, it can be distracting. With the right prompts and specific templates, an AI tool can ensure that the documentation looks as if it has been cast from a single mold.

- **Comprehensibility**
  The people who work with the source code must understand the documentation. It should therefore be written in clear and precise language. Sometimes, complex concepts need to be explained as clearly as possible. Another complicating factor is that the documentation is usually written in English, but not everyone involved is a native speaker. AI can provide support here, formulate the documentation, and help with translation.

6.2   Inline Documentation

Inline documentation includes all comments and documentation that are embedded directly in the source code of an application. This type of documentation helps to make the code easier to understand as you read it.

You should use inline comments if you want to explain individual sections of code. They are mainly used for complex code sections that are difficult to understand. A typical example is a code passage that isn't intuitive. If you've implemented a part of the business logic in such a way that it works but doesn't correspond to the code standard of your application, you should comment on this. If you were to adapt the code accordingly, the logic would no longer work as desired. To prevent you or another person from making this change at a later date, you can leave a comment.

Like almost everything in software development, inline documentation has both advantages and disadvantages. Depending on the context in which you use the documentation, these aspects come into play to a greater or lesser extent.

One advantage of inline documentation is that you can use it immediately and thus explain complex code directly. The main disadvantages are the reduced readability of the code and potential over-commenting.

As the disadvantages of inline documentation clearly outweigh the advantages, you should use it sparingly and only in areas that truly require explanation. You should formulate the remaining source code of your application in such a self-explanatory way that no inline comments are required.

6.2.1   Sample Inline Documentation

As a concrete example of the useful insertion of inline comments, we use a function that calculates the nth Fibonacci number. Without comments, the code of the function reads as follows:

```
const fibonacci = (n: number): number => {
  if (n < 0) {
    throw new Error("n must be a non-negative number");
  }
  if (n === 0) return 0;
  if (n === 1) return 1;
  let a = 0;
  let b = 1;
  for (let i = 2; i <= n; i++) {
    const temp = a + b;
    a = b;
    b = temp;
```

```
    }
    return b;
};
```

The only indication of the function's purpose is its name. Both the code and the abbreviated variable names don't give away at first glance what the function is actually supposed to do. So in the next step, we ask ChatGPT to insert explanatory inline comments:

> **Prompt:** Please add inline comments to the following function to explain what is happening.
>
> [code of the function]

Sure, here is the function with inline comments explaining each step:

```
const fibonacci = (n: number): number => {
  // Check if the input is a negative number and throw an
  // error if it is
  if (n < 0) {
    throw new Error("n must be a non-negative number");
  }

  // If n is 0, the Fibonacci number is 0
  if (n === 0) return 0;

  // If n is 1, the Fibonacci number is 1
  if (n === 1) return 1;

  // Initialize the first two Fibonacci numbers
  let a = 0;
  let b = 1;

  // Loop from 2 to n to calculate the nth Fibonacci number
  for (let i = 2; i <= n; i++) {
    // Calculate the next Fibonacci number
    const temp = a + b;
    // Update the previous two Fibonacci numbers
    a = b;
    b = temp;
  }

  // Return the nth Fibonacci number
  return b;
};
```

These explanatory comments, which expound on the purpose of almost every line, make the algorithm easy to understand. Lines that belong together and are explained with a comment have also been combined here by the AI tool and separated from the rest by a blank line. This type of documentation helps you create algorithms so that they are explained directly in the code for other people or yourself in the future. However, you can also use this type of AI support if you need to maintain code that you've taken over from other people. Before you laboriously work out the meaning of the code yourself, you can have it explained to you in this way and include this explanation as part of the code in the version control.

For a meaningful inline comment, the AI tool needs the source code you want to comment on and an instruction on what exactly you want to achieve. In this case, the task was clear, as the entire function was to be commented on step-by-step. However, you can also have only individual comments inserted. Here is an example of the calculation of a discounted price.

```
function calculateDiscount(price, discount) {
  const validDiscount = discount > 100 ? 100 : discount;

  const finalPrice = price - (price * (validDiscount / 100));

  return finalPrice;
}
```

You can ask ChatGPT to explain to you what the line containing validDiscount is supposed to do and also insert a comment.

Prompt: What does the line with validDiscount do? Please add an inline comment.

[code of the function]

The line with validDiscount is checking if the discount value is greater than 100. If it is, it sets validDiscount to 100, effectively capping the discount at 100%. Otherwise, it keeps the discount as is. This ensures that the discount doesn't exceed 100%, which wouldn't make sense in a typical pricing scenario.

```
function calculateDiscount(price, discount) {
  // Ensure the discount does not exceed 100%
  // by capping it at 100
  const validDiscount = discount > 100 ? 100 : discount;
  ...
}
```

[more details]

You'll have to deal with such cases much more frequently than with commenting on each individual line. In this case, however, you also need to give your AI tool more information about what exactly you expect.

You should use inline documentation sparingly, as it doesn't always improve the readability of the source code. Function and class documentation is used much more frequently.

6.3   Function and Class Documentation

In function and class documentation, you view functions or classes as a unit and describe them. In both cases, you explain the purpose of the respective structure.

You use function and class documentation to give developers an overview of the structure. This documentation provides a compact description of all the important aspects you need to know about the structure to use it effectively. The requirements for function and class documentation differ in detail.

The most important information you should generally provide includes the following:

- **Brief description**
  Give a one-sentence explanation that captures the purpose of the structure and mentions what you can do with it and why it exists.
- **Detailed description**
  Add details in addition to the aspects of the brief description.
- **Examples**
  You can specify how a function or class is to be used and add a short code example.
- **Known errors or to-dos**
  If there are known restrictions, errors or things that still need to be done, the documentation block is a good place to document this.

For a class, you can also specify dependencies or information about the main author.

In the case of function documentation, in addition to the preceding information, you also describe the parameters, the return value, and any exceptions the function can throw.

There are standards for class and function documentation in the various programming languages. In the case of TypeScript and JavaScript, this is JavaScript Documentation (JSDoc).

6.3.1   Benefits and Drawbacks of Function and Class Documentation

Compared to inline documentation, function and class documentation provides better context as it's applied to an entire code block. This also makes it clearer and provides a good overview. However, good function and class documentation is time consuming and carries the risk of quickly becoming outdated. This is exactly where AI tools can help.

You'll find this type of documentation much more frequently than inline documentation. Especially in areas where the source code is made available to other teams or projects in the form of libraries, there is no way around this type of documentation. Developer manuals are also often generated automatically from the comment blocks. To keep the effort to a minimum, often only the external interfaces are documented. Many developers don't provide any detailed documentation for internal or private classes, methods, and functions.

6.3.2   Sample Function and Class Documentation

The `Person` TypeScript class serves as a concrete example of function and class documentation. Here, you can see not only how you can document a class but also what this can look like for a function, as it makes hardly any difference whether you document a function or a method of a class.

The source code of the class reads as follows:

```
class Person {
  private name: string;
  private age: number;

  constructor(name: string, age: number) {
    this.name = name;
    this.age = age;
  }

  greet(): string {
    return `Hello, my name is ${this.name} and
            I am ${this.age} years old.`;
  }
}
```

With this source code, you can ask the AI tool of your choice, such as ChatGPT, to add the documentation, which you can do via the following prompt:

Prompt: Please add the missing class and method documentation to the following class:

[code of the class]

Here's the `Person` class with added JSDoc comments for class and method documentation:

```
/**
 * The `Person` class represents a person with a name and age.
 */
class Person {
  private name: string;
  private age: number;

  /**
   * Creates an instance of the `Person` class.
   *
   * @param {string} name - The name of the person.
   * @param {number} age - The age of the person.
   */
  constructor(name: string, age: number) {
    this.name = name;
    this.age = age;
  }

  /**
   * Returns a greeting message including the person's name and
   * age.
   *
   * @returns {string} A string containing the greeting
   * message.
   */
  greet(): string {
    return `Hello, my name is ${this.name} and
            I am ${this.age} years old.`;
  }
}
```

Summary of the documentation:

- Class Level: Describes the purpose of the Person class.
- Constructor: Documents the parameters required to create a Person object.
- Method: Describes the greet() method and its return value.

Even this simple example shows that you can save yourself a lot of work by letting the AI tool create the class and function documentation for you. To ensure that prompts like this work smoothly, you should make sure your code is cleanly structured. The naming of parameters also helps AI generate the documentation. If you also use a type-safe language such as TypeScript, you have the option of specifying the types of the parameters and the return value. The more information you give the AI tool, the better the result will be.

As always when working with AI, you should at least skim the generated source code and make sure that the comment blocks match the functions. However, AI tools such as ChatGPT can also derive very well from the code and generate the documentation, even if the code isn't really cleanly written, as the following example of a binary search shows:

```
function s(a, b) {
    let l = 0, r = arr.length - 1;
    while (l <= r) {
        const m = Math.floor((l + r) / 2);
        if (arr[m] === target) return m;
        if (arr[m] < target) l = m + 1;
        else r = m - 1;
    }
    return -1;
}
```

You can use the following prompt to generate a comment block for the function:

Prompt: Could you please write a documentation for the following function:

[code of the function]

Certainly! Here's a detailed documentation for the s function:

```
/**
 * Performs a binary search on a sorted array to find the index of a target
element.
 *
 * ...more comments here...
```

[explanation: more details]

Here, too, the AI tool has done a great job and created a very detailed commentary block. It contains both a description of the parameters and the return value as well as examples of use and a detailed description of the function. In addition, the tool has added a note that the input array must be sorted; otherwise, the behavior is undefined.

6.4 API Documentation

The last level of documentation we want to introduce in this chapter is API documentation and, specifically, the documentation of a web service's REST API. When developing web applications, you usually have a natural break between frontend and backend.

The endpoints of the backend provide read and write interfaces for the frontend. Front-end and backend are often implemented by different teams, sometimes even by different companies. For this reason, clear and proper communication is of great importance at this point.

The OpenAPI Specification, also known under its former name, Swagger, is widely used for documenting REST APIs. Depending on the platform used, you describe the interfaces of your application in a separate document in YAML Ain't Markup Language (YAML), JavaScript Object Notation (JSON), or directly in the code as a comment or using a decorator.

6.4.1 Sample API Documentation

To create API documentation, we use an existing TypeScript application based on the Express framework. There is an extension for Express available called `swagger-ui-express`, which provides a graphical interface with the API documentation via a given path. In our example, the extension reads its configuration from a Swagger file, which is initially still empty, and makes the documentation available via the /api path.

The application has an endpoint, namely the /users path, which is used to deliver user data in JSON format. The data originates from a UserService class and represents instances of the User class. The code of the initial file of the application looks as follows:

```
import { readFileSync } from "node:fs";
import express from "express";
import swaggerUi from "swagger-ui-express";
import * as YAML from "yaml";
import UserService from "./user.service.js";

const app = express();

const yamlSpec = readFileSync("./swagger.yaml", "utf8");
const swaggerDocument = YAML.parse(yamlSpec);

app.use(
  "/api",
  swaggerUi.serve,
  swaggerUi.setup(swaggerDocument)
);

const userService = new UserService();
app.get("/users", (request, response) => {
  response.json(userService.getAll());
});
```

```
app.listen(3000, () => {
  console.log("Server is running on port 3000");
});
```

In this case, we use the GitHub Copilot extension for Visual Studio (VS Code) to generate the API documentation. When you open the swagger.yaml file in your development environment, you can use the chat feature to issue the following prompt:

Prompt: Please create an API documentation for the express application in #file: index.ts using the user class in #file:user.ts. Please include example data.

Certainly! Here's an example of how you can create an API documentation using Swagger YAML in the swagger.yaml file:

```yaml
openapi: 3.0.0
info:
  title: User API
  version: 1.0.0
servers:
  - url: http://localhost:3000
paths:
  /api/users:
    get:
      summary: Get all users
      responses:
        "200":
          description: OK
          content:
            application/json:
              schema:
                type: array
                items:
                  $ref: "#/components/schemas/User"
components:
  schemas:
    User:
      type: object
      properties:
        name:
          type: string
          example: John Doe
        email:
          type: string
          example: johndoe@example.com
```

```
password:
  type: string
  example: password123
```

The context of API documentation is much more extensive than that of a simple function or class. The API is the entry point for client-server communication. The associated source code comprises not just a single file, but usually multiple files. In the example, these are the initial index.ts file and the user.ts file with the definition of the structure. You must make these files available to the AI tool to obtain a result that makes sense.

Apart from one small error, Copilot delivers exactly the result we expect. You only need to change the /api/users path to /users. Other than that, both the general information and the documentation for the endpoint and the data structure with the required sample data are correct. If you apply the code and reload your application, you can access the Swagger UI via *http://localhost:3000/api* and test your endpoint from there.

Figure 6.1 View of the Swagger UI

Click the **Try it out** button, and then **Execute** to send the request to the server and test your endpoint. Due to the configuration, the graphical interface not only provides information about the path and HTTP method but also about the structure of the response and the associated sample data.

6.5   Outdated Documentation

There are few things worse than outdated documentation. If you've ever had the problem in an application that you couldn't rely on the documentation (for example, the comment block for a function), you'll invest more time in research in future and, if in doubt, take an additional look at the code. The implementation of an extension will therefore take more time.

How does outdated documentation arise? Usually, this doesn't happen because malicious people want to lead others up the garden path, but mostly it's due to time pressure or because the documentation update task was simply forgotten. What remains, however, is the loss of trust, and thus the documentation loses considerable value.

If you work on updating a function yourself, you can get help from AI tools and have the documentation updated automatically. The `fetchUserData` function serves as an example of such a workflow. It was originally implemented in such a way that you pass it the ID of a user and a callback function. The function itself loads the data from a web server and then calls the callback function. The function is now implemented as an `async` function thanks to a modification. It only accepts the user's ID and returns a `Promise` object with the loaded user data. However, the function documentation still refers to the old implementation. In the code it looks as follows:

```
/**
 * Fetches user data from the server using the Fetch API.
 *
 * This function sends a GET request to the server to retrieve
 * user data based on the provided user ID. The response is
 * expected to be in JSON format.
 *
 * @param {number} userId The ID of the user whose data
 *                        is to be fetched.
 * @param {function} callback A callback function that will
 *                        be called with the user data.
 *                        The callback should accept two
 *                        parameters: an error (if any) or
 *                        null if the request was successful
 *                        and the user data.
 *
```

```
 * @return {void}
 */
async function fetchUserData(userId: number) {
  const response = await fetch(`/api/users/${userId}`);
  if (!response.ok) {
    throw new Error('Failed to fetch user data');
  }
  return response.json();
}
```

The description of the function is still correct, but the explanation of the parameters and the return value is incorrect. To resolve this issue, you can, for example, use VS Code's GitHub Copilot integration again and submit the following prompt:

Prompt: Please update the documentation for the `fetchUserData` function.

```
/**
 * Fetches user data from the server.
 *
 * @param userId - The ID of the user to fetch data for.
 * @returns A promise that resolves to the user data.
 * @throws An error if the request fails.
 */
async function fetchUserData(userId: number) {
  const response = await fetch(`/api/users/${userId}`);
  if (!response.ok) {
    throw new Error("Failed to fetch user data");
  }
  return response.json();
}
```

What is immediately noticeable here is that Copilot has intervened in the structure of the commentary. The original, detailed description of the function has been removed, and the parameter types have also been omitted. Overall, the result is correct, but some information has been lost. This is a good example of how an AI tool with a prompt that is too simple doesn't necessarily improve the situation. You can modify the prompt so that the result is more like the original variant. However, this is only worthwhile for more extensive source code, as in this case, you would invest so much time in the prompt that you might as well write the documentation yourself.

An alternative is to use another tool such as ChatGPT. The result looks different with the same prompt:

Prompt: Please update the documentation for the following function:

[code of the function]

Here's the updated documentation for the `fetchUserData` function:

```
/**
 * Fetches user data from the server using the Fetch API.
 *
 * This asynchronous function sends a GET request to the server
 * to retrieve user data based on the provided user ID. The
 * response is expected to be in JSON format and is returned as
 * a Promise that resolves to the user data.
 *
 * @param {number} userId - The ID of the user whose data is to
 *                          be fetched.
 *
 * @throws {Error} Will throw an error if the network request
 *                 fails or if the server returns a non-OK
 *                 status.
 *
 * @returns {Promise<Object>} A Promise that resolves to the
 *                            user data object when the request
 *                            is successful.
 *
 * @example
 * fetchUserData(1)
 *    .then(userData => console.log(userData))
 *    .catch(error => console.error('Error fetching user data:',
 *                  error));
 */
async function fetchUserData(userId: number) {
  const response = await fetch(`/api/users/${userId}`);
  if (!response.ok) {
    throw new Error('Failed to fetch user data');
  }
  return response.json();
}
```

[key updates: more details]

ChatGPT hasn't deleted any information but rather has adapted the documentation of the parameters and added an example of how to use the function. You'll also receive a list of the changes made, including the associated explanations.

When using AI tools, you also have to weigh the convenience offered by a tool such as GitHub Copilot, which allows you to modify the code directly in the development environment, against the quality of another tool, which may require you to paste the source code into another command prompt.

6.6   Conclusion

Documentation is an important part of software development, but it's not as much fun as writing production code for new features. For this reason, documentation is often either not prepared or only to an inadequate extent. The situation becomes even worse if the documentation hasn't been updated when a change is made to the code, as this can lead to a loss of trust among the developers involved. In these situations, you can use AI tools to help you generate documentation and keep it up to date. With the different types of documentation, you may need to adjust your prompts slightly and provide the AI with sufficient context.

If you're just starting to generate documentation, you should test different tools. The various tools differ in the way they are used and also in the quality of the results. Select the tool that best suits you, your application, and your workflow.

> **Changing Documentation Automatically**
>
> In Chapter 10, we look at the question of the extent to which it's possible to automatically add comments to an entire collection of files in a project or to switch the documentation from German to English. But let me say this first: the smaller the code files, the better your results.

Chapter 7
Databases

Databases may seem an unusual topic for a book that deals with coding. But SQL is also code, and AI tools are excellent at compiling SQL commands. In addition, AI is also a great help in other areas, such as the design of databases or the administration of database management systems (DBMSs).

In this chapter, we'll use a wide range of examples to illustrate the use of AI tools in the design or modification of databases, the composition of SQL commands, administration, and finally the development of client code.

DBMS-Specific Answers

Relational database management systems (RDBMSs), such as Microsoft SQL Server, MySQL, Oracle, and PostgreSQL have many things in common, but they differ significantly in the details. Unless you have a very basic question, you should always specify in the prompt which DBMS your question relates to! Otherwise, you may receive answers that don't apply to your DBMS at all.

You'll have the best experience with ChatGPT and others if there is a lot of training material for your DBMS available on the internet. With relational systems, this applies in particular to MySQL/MariaDB, PostgreSQL, SQLite, and SQL Server; with NoSQL products, the very popular programs MongoDB and Redis have an advantage.

7.1 Database Design

As we're going to show you in the following examples, AI tools can save you a lot of work when designing databases and formulating the necessary CREATE TABLE commands. As is so often the case, the more precisely you formulate your wishes and the more clearly you can assess which parts of the AI suggestions actually correspond to your requests, the better the AI support will work. If you're not satisfied, you can make improvements through targeted prompts.

7.1.1 Example: Database for an Online Store

Let's assume you want to implement an online store. You pursue a dynamic pricing policy; that is, you want to adjust product prices at any time. So, ask ChatGPT for an initial database design.

> **Prompt:** I want to create a database for an online store. The prices of my products will change frequently (sometimes daily). Can you help me with logical modeling?

> Absolutely, I can help you design a logical model for your online store database. Given that the prices of your products change frequently, we'll need to ensure that the database structure supports efficient updates and historical tracking of price changes.
>
> Here's a basic logical model for your online store database.
>
> **Entities and attributes:**
>
> - Users
>   - UserID (Primary Key)
>   - Username
>   - PasswordHash
>   - Email
>   - DateRegistered
> - Categories
>   - CategoryID (Primary Key)
>   - CategoryName
>   - Description
> - ...
>
> **Relationships:**
>
> - *Users* can place multiple *Orders*.
> - Each *Order* can have multiple *OrderDetails*.
> - ...
>
> [many more details]

There is little to criticize about the database design suggested by ChatGPT. The organization of the data makes sense, and the number of entities is also reasonable for this design phase. For further discussion of the database design, an entity relationship (ER) diagram would be a good thing to have. Confident as ever, ChatGPT introduces the corresponding request with *Sure*, but then has to admit that it can't draw diagrams. (We're very curious to see when AI tools will be able to do this.)

> **Prompt:** Can you draw an ER diagram for the database?

Sure, I can describe how the ER diagram for this database would look. Given that I'm unable to directly draw images, I'll provide a detailed description of how you can create the ER diagram. You can then use tools like draw.io, Lucidchart, or any ER diagram tool of your choice to visualize it.

[many more details]

ChatGPT essentially repeats the output of the previous prompt and gives (a few) tips on how this data could be visualized.

But you can outsmart ChatGPT with a trick! There are various tools on the internet for drawing diagrams that expect statements similar to a programming language. This will take ChatGPT back to its text model domain. (You might wonder why ChatGPT doesn't make such a suggestion of its own accord.)

So, ask ChatGPT to create code to draw the ER diagram with Mermaid. Mermaid (*https://mermaid.js.org*) is a great JavaScript library for generating charts and diagrams:

Prompt: Can you provide code to draw an ER diagram with Mermaid?

Sure! Mermaid is a great tool for creating diagrams, including ER diagrams, with code. Below is the Mermaid code to generate the ER diagram for your online store database:

```
erDiagram
    USERS {
        int UserID PK
        string Username
        string PasswordHash
        string Email
        date DateRegistered
    }
    ...
    USERS      ||--o{  ORDERS : "places"
    CATEGORIES ||--o{  PRODUCTS : "contains"
    PRODUCTS   ||--o{  PRODUCT_PRICES : "has"
    ORDERS     ||--o{  ORDER_DETAILS : "includes"
    PRODUCTS   ||--o{  ORDER_DETAILS : "part of"
```

To use this code:

- Copy the Mermaid code above.
- Paste it into a Mermaid live editor.
- The diagram will be rendered automatically.

This Mermaid code defines the entities (tables), their attributes, primary keys (PK), and foreign keys (FK), along with the relationships between them. The relationships are depicted with the cardinalities: ||--o{ representing a one-to-many relationship.

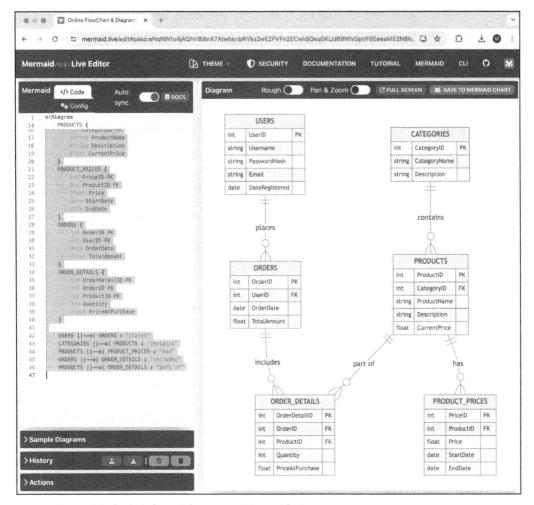

Figure 7.1 The Database Schema as a Mermaid Diagram

This approach works extremely well in a general sense; however, the visual design is certainly open to debate. If required, you can use options to adapt colors, fonts, and design themes to your wishes. If you're not at all happy with Mermaid, you can alternatively ask ChatGPT to formulate the code of the diagram in the syntax of PlantUML, Graphviz, or TikZ.

From the perspective of database theory, ChatGPT can be criticized for not providing an ER diagram at all, but rather a visual representation of the database schema. (An ER diagram is only a preliminary stage. It contains fewer implementation details, for example, no data types such as int or date. But "real" ER diagrams are much clearer.)

Let's assume you're basically satisfied with the design. The next step is to create the database using CREATE-TABLE commands. At this point at the latest, you must decide on a specific DBMS (here: Microsoft SQL Server), and you can also add further entities or columns to the schema, formulate requests for naming the tables and columns, and so on.

> **Prompt:** I would like to implement the database using Microsoft SQL Server. Can you help me with CREATE TABLE commands? Also suggest indexes.
>
> Please add some additional columns:
>
> - Customers: first name, last name, address (street, city, etc.)
> - Products: URL for product description
> - Product prices: VAT rate
>
> Use lowercase names, plural for tables, singular for columns.

Certainly! Below are the CREATE TABLE commands for your database in Microsoft SQL Server, including the additional columns you requested. I've also included suggestions for indexes to optimize query performance.

```
CREATE TABLE users (
    userid INT PRIMARY KEY IDENTITY(1,1),
    username NVARCHAR(50) NOT NULL,
    passwordhash NVARCHAR(255) NOT NULL,
    email NVARCHAR(100) NOT NULL,
    dateregistered DATE NOT NULL,
    firstname NVARCHAR(50),
    lastname NVARCHAR(50),
    street NVARCHAR(100),
    city NVARCHAR(50),
    state NVARCHAR(50),
    zip NVARCHAR(20),
    country NVARCHAR(50)
);
...


-- Indexes
CREATE INDEX idx_users_email ON users(email);
CREATE INDEX idx_categories_categoryname
        ON categories(categoryname);
CREATE INDEX idx_products_categoryid
        ON products(categoryid);
...
```

[explanation: more details]

Warning

It should be clear that even with AI support, you won't be able to quickly design a new database in half an hour. An inadequately thought-out schema will cause major problems as a result. This isn't the fault of the AI tool you're using, but of your insufficiently precise description of the problem.

Of course, this also applies to the example presented here: the customers table is missing columns for authentication in the web store (login). The orders table lacks information on how payment is made (invoice, PayPal, etc.). The shipping of products (including the calculation of shipping costs) is also not modeled. You usually want to display images with each product. This requires one or two additional tables and considerations as to whether the images should be saved as files or directly in the database. ChatGPT and others are happy to fulfill all these additional requests, but you need to formulate them!

Long story short is that AI tools help you avoid tedious typing. However, you still need to think for yourself about what tasks your database should perform and how best to implement these functions.

7.1.2   Database Management System Selection

In the previous example, we assumed that Microsoft's SQL Server would be used. However, selecting the optimal DBMS is often a topic in itself. AI tools can also help with this, albeit with limitations. In our tests, we've noticed a tendency for AI tools to be reluctant to provide a clear recommendation. Instead, you get a long list of arguments for and against individual products. ChatGPT can't and won't take the decision away from you, but at least it provides you with objective arguments.

The following prompt refers to the online store database discussed earlier.

Prompt: A friend suggested that I use a NoSQL database instead of SQL Server. He mentioned MongoDB. Would this make sense?

Using a NoSQL database like MongoDB can make sense depending on your specific use case and requirements. Here are some considerations to help you decide whether MongoDB or another NoSQL database is suitable for your online store:

Advantages of Using MongoDB (NoSQL): Schema flexibility, horizontal scalability, document-oriented storage, performance [more details]

Potential Drawbacks of Using MongoDB: Lack of ACID transactions, complex queries, data duplication [more details]

Example MongoDB Schema for Your Online Store

Here is an example of how you might structure your data in MongoDB.

[more details]

Summary

- Use MongoDB if you need flexible schemas, horizontal scalability, and efficient handling of large volumes of semi-structured or unstructured data.
- Use SQL Server if your application relies on complex transactions, relationships, and requires strong ACID compliance.

Ultimately, the choice depends on your specific needs, including data structure complexity, scalability requirements, and the nature of your queries and transactions.

ChatGPT lists various advantages and disadvantages of MongoDB compared to SQL Server on approximately two pages of text (reproduced here in abbreviated form). ChatGPT also provides a suggestion for implementing four MongoDB collections to store the data.

After consulting with the team, you decide against MongoDB, but you're unsure as to whether SQL Server is the best choice.

Prompt: After consulting with my team, we decided to use an RDBMS. However, we're considering an open-source product (MariaDB, MySQL, PostgreSQL) instead. Can you recommend one of these programs?

All three of the open-source RDBMS options you mentioned (MariaDB, MySQL, PostgreSQL) are robust and widely used, but they have some differences that may make one more suitable for your specific needs. Here's a brief comparison to help you decide:

[more details]

Summary

- PostgreSQL is recommended for its advanced features and performance.
- MariaDB is a strong alternative, especially if you prefer MySQL compatibility and a fully open-source solution.
- MySQL remains a popular choice for its ease of use and extensive support.

Choose the one that best fits your team's familiarity and the specific needs of your application. If you need further assistance with setting up or configuring your chosen RDBMS, feel free to ask!

7.1.3   Example: Cloud Shopping List

You may be familiar with apps family members use to organize their shopping. If you notice that you're missing milk, flour, or butter, you write the item on a shopping to-do list. Whoever goes shopping next will see the jointly created list and can tick it off again while shopping.

Prompt: We want to build a smartphone app with a cloud backend. In the app, all members of a group (typically a family) can add shopping items. Whoever does the next shopping will see all the items that are currently missing at home. While shopping, he/she can uncheck all items in the shopping cart.

I am responsible for the database backend, which will be implemented using MongoDB. Can you suggest a database schema to handle the shopping list and connect the members of a group?

Certainly! Below is a suggested schema for a MongoDB database to handle a shopping list app where group members can add and manage items.

User collection: Each user has an account and belongs to one or more groups.

```
{
  "_id": "ObjectId",          // Unique identifier for the user
  "username": "string",       // Username of the user
  "email": "string",          // Email address
  "password": "string",       // Hashed password
  "groups": ["ObjectId"]      // Array of group IDs the user
}                             // belongs to
```

Group collection: Each group contains multiple users and a list of shopping items.

```
{
  "_id": "ObjectId",          // Unique identifier for the group
  "name": "string",           // Name of the group (e.g.,
                              // Family, Friends)
  "members": ["ObjectId"],    // Array of user IDs who are
                              // members of this group

  "shoppingList": [
    {
      "itemId": "ObjectId",   // Unique identifier for the item
      "name": "string",       // Name of the item
      "quantity": "string",   // Quantity (2 liters, 1 pack ...)
      "addedBy": "ObjectId",  // User ID who added the item
      "status": "string"      // Status of the item (pending,
    }                         // purchased ...)
  ]
}
```

Example Documents: [more details]

Example Queries: [more details]

Key Points:

- **User Authentication:** Store user credentials and provide authentication mechanisms.

- **Group Management:** Each user can belong to multiple groups, and each group can have multiple members.
- **Shopping List Items:** Items in the shopping list have statuses to track whether they are pending or purchased.

ChatGPT's answer goes beyond the actual question. Not only does the AI tool suggest a useful database schema, it also provides concrete examples of how to apply the schema and perform basic database operations (adding items to the shopping list, etc.).

Our earlier warning that you should critically scrutinize ChatGPT's proposed solution naturally also applies here. You probably have detailed requests for the visual and functional design of the app that require corresponding changes to the database schema. But as a starting point for an initial prototype, the answer provided by ChatGPT is perfectly adequate.

7.1.4 Changing the Database Schema

In real life, you rarely have the opportunity to develop a completely new database schema from scratch. More often, you'll only need to expand the schema regarding a few specific details. AI tools can also help in this respect. Typically, in the first step, you transfer the current schema of the database (i.e., all underlying CREATE TABLE commands) or make it available in an editor tab. (If required, simply ask ChatGPT how to create a schema file. You can find a corresponding example in Section 7.3.)

For some questions, it's sufficient to specify only the definition of a single table. The following question is about how large a text column must be so that IPv6 addresses can be stored without errors in all special cases.

> **Prompt:** MySQL: We use the `logins` table to store IPv4 addresses.
>
> ```
> CREATE TABLE logins ( ...
>   ip VARCHAR(15) DEFAULT NULL,
>   ...)
> ```
>
> Now we have to deal with IPv6 addresses as well. How do we change the `ip` column? Could you provide an ALTER TABLE command?

```
ALTER TABLE logins MODIFY COLUMN ip VARCHAR(45) DEFAULT NULL;
```

This change sets the length of the `ip` column to 45 characters, which is sufficient to store both IPv4 and IPv6 addresses, including the full expanded form of an IPv6 address.

The starting point for the second example is a moderately complex database for storing customer data, orders, and invoices. The schema file comprises approximately 300 lines and contains 10 linked files.

You're working in VS Code, and the schema file is located in your project directory. You want to change the structure of the database. First, you need to make sure that your AI tool understands the structure of the database at all.

> **Prompt:** @workspace Can you explain the content of `accounting-schema.sql`?

> The `accounting-schema.sql` file appears to be a MySQL database schema dump for a database named `accounting`. It contains the SQL commands to recreate the structure of the database, including the creation of tables and their relationships. Here are some of the tables defined in the schema . . .
>
> [more details]

Once this test has been completed to your satisfaction, you can formulate what you actually want. Only the contact details of one person can currently be saved in the `customers` table. For some of your customers, however, you want to save several persons.

> **Prompt:** OK. The `customers` table currently contains contact information for only one person. For some customers, we need to store data for several contact persons (first name, last name, email, phone number). Can you suggest a schema change? Existing data must be preserved.

> To accommodate multiple contact persons for a single customer, you can create a new table called `contactPersons`. This table will have a foreign key linking it to the `customers` table. Here's the SQL to create this new table:
> ```sql
> CREATE TABLE contactPersons (
>   id         INT(11) NOT NULL AUTO_INCREMENT,
>   customerId INT(11) NOT NULL,
>   firstName  VARCHAR(100) DEFAULT NULL,
>   lastName   VARCHAR(100) DEFAULT NULL,
>   email      VARCHAR(100) NOT NULL,
>   telNo      VARCHAR(100) DEFAULT NULL,
>   PRIMARY KEY (id),
>   FOREIGN KEY (customerId) REFERENCES customers(id)
> );
> ```
> In this new schema, each customer can have multiple contact persons. The `customerId` field in the `contactPersons` table links each contact person to a customer in the `customers` table.

To preserve existing data, you can migrate the contact person data from the `customers` table to the `contactPersons` table with a SQL command like this:

```
INSERT INTO contactPersons (customerId, firstName,
                            lastName, email, telNo)
SELECT id, firstName, lastName, email, telNo
FROM customers;
```

After the data migration, you may want to keep the contact person fields in the `customers` table for a while for backward compatibility. However, you should plan to eventually remove these fields and update any code that references them to use the new `contactPersons` table instead.

We ran the example with GitHub Copilot, and the tool really excelled in this case: not only does it suggest a sensible schema change (outsourcing the contact data to a separate table), it also adopts the column names and data types from the original `customers` table, names the new table in the same camelCase notation that was used for the previous tables, and remembers the new foreign key rule for linking the new table to the existing `customers` table. The suggestions for data migration make sense as well and work without a problem.

Of course, it's still up to you whether you want to carry out the database conversion as suggested or whether changes are still necessary. If you have multiple contacts, you may want to mark one of them as the main contact, in which case, you'll need an additional column.

7.2   SQL Commands

Structured Query Language (SQL) is used to extract data from databases and to insert, change, or delete new data. In addition, there are countless SQL commands to create or change new databases and to perform administrative tasks.

In this section, we'll give you some examples of how ChatGPT can help you develop correct SQL commands. The syntax of SQL is defined by standards; SQL therefore functions mostly identically in all relational DBMSs. However, almost all DBMSs are characterized by manufacturer-specific deviations from the standard. In this respect, it's important that you always specify which program you're referring to.

7.2.1   Date Calculation

It happens time and again that you have to process dates or times. But how can you determine the orders of the past six months or the logging entries in the past three hours? How do you set a reminder for a meeting that is going to take place in a week's time? The functions for time and date calculations vary depending on the DBMS.

Prompt: MySQL: How do I get the date six months before now?

In MySQL, you can use the `DATE_SUB` function to get the date six months before the current date. Here's how you can do it:

```
SELECT DATE_SUB(NOW(), INTERVAL 6 MONTH) AS six_months_ago;
```

This command will return the date and time exactly six months before the current date and time. If you only need the date portion without the time, you can use `CURDATE()` instead of `NOW()`:

```
SELECT DATE_SUB(CURDATE(), INTERVAL 6 MONTH) AS six_months_ago;
```

7.2.2 Troubleshooting

The particularities of SQL syntax can sometimes drive you mad. AI tools succeed in solving the puzzles in many cases. In simple cases, you don't even have to tell the AI tool the schema of the database. For queries with links across multiple tables, ChatGPT and others can give more precise advice if they know the data types of the table columns.

Prompt: I want to find categories with more than 10 titles in my book library. However, my query fails (syntax error). Why is this?

```
SELECT c.name AS categoryName, COUNT(*) AS cnt
FROM books b JOIN categories c ON c.id = b.catId
GROUP BY c.id
WHERE cnt >= 10
```

The issue with your query is the placement of the `WHERE` clause. In SQL, the `WHERE` clause is used before the `GROUP BY` clause. If you want to filter the results of a `GROUP BY` operation, you should use the `HAVING` clause. Here's how you can fix your query:

```
SELECT c.name AS categoryName, COUNT(*) AS cnt
FROM books b JOIN categories c ON c.id = b.catId
GROUP BY c.id
HAVING cnt >= 10
```

We executed the second example in VS Code using GitHub Copilot. The AI tool had access to the database schema file and even stated in its response that it had used this information.

Prompt: In my accounting database, I want to create a table of all customers with the total number of invoices. However, new customers who haven't received an invoice yet are missing from the result. What's wrong with my query?

```
SELECT company, COUNT(*)
FROM customers c JOIN invoices i on i.customerId = c.id
GROUP BY c.id
```

The issue with your query is that you're using an INNER JOIN which only returns rows where there is a match in both tables. If a customer doesn't have an invoice yet, they won't appear in the result set.

To include customers who haven't received an invoice yet, you should use a LEFT JOIN. This type of join returns all the rows from the left table (customers), and the matched rows from the right table (invoices). If there is no match, the result is NULL on the right side.

Here's how you can modify your query:

```
SELECT c.company, COUNT(i.invoiceId)
FROM customers c
LEFT JOIN invoices i ON i.customerId = c.id
GROUP BY c.id, c.company;
```

This query will return all customers, and for those without invoices, it will show a count of 0.

GitHub Copilot has correctly identified the cause of the error and has basically proposed a good solution. What is annoying here is that it has introduced a new error that could have been avoided: in COUNT(i.invoiceId), the AI tool assumes that the ID column of the invoices table is named invoiceId. This may seem obvious, but in this specific case, the column is simply called id. This information is clear from the schema file.

7.2.3   WordPress Image Search

Recently, we wanted to add alt attributes to images in a WordPress database for an additional image description. We knew that WordPress uses MySQL or MariaDB as a DBMS, but we had never looked into the structure of this database before. ChatGPT came in handy.

Prompt: WordPress/MySQL: How can I get the newest 10 uploaded images with URL, description, and alt text from the WordPress database?

To retrieve the newest 10 uploaded images, you'll need to execute a SQL query on the WordPress MySQL database. WordPress stores information about uploaded images (and all media library items) in the wp_posts table and the associated metadata in the wp_postmeta table. The URL of an image, its description, and its alt text are stored as metadata.

The following query is an example of how to retrieve this information. This query assumes you're using the default WordPress table prefix (wp_). If you have a different table prefix, you'll need to replace wp_ with your custom prefix in the query:

```sql
SELECT p.ID,
       p.guid AS 'URL',
       p.post_content AS 'Description',
       pm.meta_value AS 'Alt Text'
FROM  wp_posts p
LEFT JOIN wp_postmeta pm ON p.ID = pm.post_id
       AND pm.meta_key = '_wp_attachment_image_alt'
WHERE p.post_type = 'attachment'
  AND p.post_mime_type LIKE 'image/%'
ORDER BY p.post_date DESC
LIMIT 10;
```

Here's a breakdown of what this query does ... [more details]

Keep in mind that the guid field is used here to retrieve the URL, which, in general, shouldn't be used for direct URLs to media files due to potential issues with changing domains or migration. A more reliable method to get the direct URL is to use WordPress functions like wp_get_attachment_url() within the WordPress framework, but for direct SQL queries, the guid is commonly used for simplicity.

The query is helpful because it's a good starting point to understand the WordPress database setup. In a second step, we now want to determine only those images for which the alt attribute is missing or empty.

Prompt: OK. Instead of p.post_content, please include p.post_excerpt for the image title. Also, I want only images that don't have an alt text.

ChatGPT provides a new query that is more suitable for our purposes (not printed here). The next step is about how we can add an alt attribute to a given image.

Prompt: Given an ID from wp_posts, what would an INSERT statement for wp_postmeta look like to set the alt text for an image?

To insert an alt text for an image in the wp_postmeta table, given an image's ID from wp_posts, you would use an INSERT statement in SQL. The image's ID is used as the post_id in the wp_postmeta table, and the alt text is stored with a meta_key of _wp_attachment_image_alt. Here's how the INSERT statement would look:

```sql
INSERT INTO wp_postmeta (post_id, meta_key, meta_value)
VALUES (<IMAGE_ID>,
```

```
        '_wp_attachment_image_alt',
        'Your alt text here');
```

[more details]

For our purposes, the information provided was sufficient to develop a Python script that searches for images without an image description and subsequently adds an image description to them. Perhaps you're wondering where the picture description comes from. We used an API account at OpenAI, uploaded the images there, and asked for an image description. The procedure worked amazingly well. However, the code has nothing to do with the content of this book, which is why we won't explain this example further here.

7.2.4 Complex Query with Subquery or Self Join

The starting point for the following example is a database from a cloud service provider. The monthly subscription fee is determined by the subscriptionPrice column. To determine the expected total monthly revenue, the basePrice of each customer's last order must be totaled. The difficult aspect from an SQL point of view is that not all data records from the orders table are taken into account, but only the last order of each customer.

To enable ChatGPT to create a suitable SELECT query, you must inform the AI tool about the structure of the database. The easiest way to do this is to specify the CREATE-TABLE commands required to create the table. (With MySQL and MariaDB, you can determine these commands using SHOW CREATE TABLE. Another option is to create a backup of the schema definition.) In this example, we've only specified the columns relevant to the example and eliminated all other columns.

Prompt: MySQL database: I have a database with two tables customers and orders:

```
CREATE TABLE customers (
  id                INT NOT NULL AUTO_INCREMENT,
  company           VARCHAR(100) NOT NULL,
  state             VARCHAR(100) NOT NULL,
                    ON UPDATE current_timestamp(),
  PRIMARY KEY (id),
)
CREATE TABLE orders (
  id                INT NOT NULL AUTO_INCREMENT,
  customerId        INT NOT NULL,
  orderType         VARCHAR(20) NOT NULL,
  orderDate         DATE NOT NULL DEFAULT curdate(),
  subscriptionPrice DOUBLE NOT NULL DEFAULT 0,
```

```
    PRIMARY KEY (id),
    CONSTRAINT orders_ibfk_1 FOREIGN KEY (customerId)
       REFERENCES customers (id)
)
```

I need to calculate a sum over subscriptionPrice for all customers with customers.state = 'ok'. The sum should only include the newest order per customer with orders.state = 'subscription'. Can you help me with the SELECT command?

Certainly! To calculate the sum of subscriptionPrice for all customers with state = 'ok', including only the newest order per customer with orderType = 'subscription', you can use a subquery to get the newest order for each customer and then join it with the customers table. Here's how you can write the SQL query:

```
SELECT SUM(o.subscriptionPrice) AS total
FROM customers c
JOIN (
    SELECT customerId, MAX(orderDate) AS latestOrderDate
    FROM orders
    WHERE orderType = 'subscription'
    GROUP BY customerId
) latest_orders ON c.id = latest_orders.customerId
JOIN orders o ON latest_orders.customerId = o.customerId
            AND latest_orders.latestOrderDate = o.orderDate
WHERE c.state = 'ok';
```

This query works as follows:

- The subquery latest_orders retrieves the customerId and the maximum orderDate for each customer where the orderType is 'subscription'.
- This subquery result is joined with the customers table to get the customers with state = 'ok'.
- Another join is performed with the orders table to get the details of the newest order for each customer.
- Finally, the SUM function is used to calculate the total subscriptionPrice for these orders.

If you have multiple orders on the same date for a customer, this query will include all of them. If you only want one order per customer, even if there are ties, you might need a more complex query to handle tie-breaking logic.

This result is great! Not only did ChatGPT understand our intention, the AI tool also provided a pretty perfect answer and immediately pointed out the only limitation: if a customer has placed *multiple* orders in one day, they will all be taken into account. (In this example, where it's known from AUTO_INCREMENT that the id column contains ascending values, this column could be used instead of lastOrderDate to link to the last order.)

ChatGPT has used a subquery in its suggested solution. This provides relatively readable SQL code, but is sometimes slow to execute. Even if no issues are to be expected in this example, we wanted to know whether ChatGPT also provides a solution without a subquery.

Prompt: Is it possible to do this query without a subquery?

```
SELECT SUM(o1.subscriptionPrice) AS total
FROM customers c
JOIN orders o1 ON c.id = o1.customerId
LEFT JOIN orders o2
    ON o1.customerId = o2.customerId
    AND o1.orderDate < o2.orderDate
    AND o2.orderType = 'subscription'
WHERE c.state = 'ok'
  AND o1.orderType = 'subscription'
  AND o2.id IS NULL;
```

This query works as follows [more details]

This answer is also correct. Now we wanted to know which variant is preferable. However, ChatGPT doesn't want to commit itself to this. Instead, the bot provides a comprehensive discussion of the advantages and disadvantages of both query types from various points of view.

Prompt: Which query is more efficient?

The efficiency of a query can depend on various factors, including the specific database system, indexing, and data distribution. However, in general, subqueries and self-joins have different performance characteristics.

[many more details]

7.2.5   Formatting an SQL Command

In our coding practice, we often try out a new SQL command interactively before transferring the final command to a code file later. In these experiments, we don't bother to format the SQL command in a legible way: we write the entire command in lowercase letters, don't use any line breaks, and so on.

If everything then works as expected, it's important for us to format the code properly using SQL keywords in uppercase letters, meaningful line breaks, and so on. This works very well in VS Code, where you mark the SQL code and then ask for a legible formatting via Copilot chat:

> **Prompt:** Please capitalize SQL keywords and insert line breaks.

Alternatively, you can copy the SQL code into a chat window. We've done this here to better illustrate the concept in the book:

> **Prompt:** Please reformat the following SQL command and capitalize all keywords.
>
> ```
> create table log (id integer primary key, url text,
> ip_address text, request_time datetime);
> ```

```
CREATE TABLE log (
    id INTEGER PRIMARY KEY,
    url TEXT,
    ip_address TEXT,
    request_time DATETIME
);
```

The code would be even easier to read with a few spaces between the column names and the data types, so that the data types are aligned in one column.

> **Prompt:** Great. Could you also add spaces between the column names and their types to align the types vertically?

```
CREATE TABLE log (
    id            INTEGER PRIMARY KEY,
    url           TEXT,
    ip_address    TEXT,
    request_time  DATETIME
);
```

Of course, this is only worthwhile for longer commands, as otherwise the prompt input takes more time than manually reformatting the code.

7.3 Administration

AI tools also help with the administration of DBMSs. Whether it's more trivial tasks such as setting up new databases or performing a backup, or whether you need to set up and configure a DBMS from scratch, ChatGPT and others are there to help you with tips.

In our experience, the quality of such instructions is usually good in simple cases. If, on the other hand, you need real specialist knowledge where a lengthy internet search will only yield a few and possibly contradictory pieces of advice, you should also treat AI answers with a great deal of caution. Section 7.3.4 provides an example of this.

7.3.1 PostgreSQL: User Administration

The setup of new databases always involves the same steps: create database, create users, and grant the relevant privileges to the users. In this respect, AI tools can save you a lot of typing.

Prompt: PostgreSQL: I want to create a new database and set up three users: one with full administrative privileges, one for database use (basically SELECT, INSERT, UPDATE, DELETE), and one for an automated backup system. Can you help me with the SQL commands I need?

Certainly! Below are the SQL commands needed to create a new database and set up three users with the specified privileges in PostgreSQL [more details]

- Connect to PostgreSQL as a superuser (usually postgres).

  ```
  psql -U postgres
  ```

- Create the database:

  ```
  CREATE DATABASE your_database_name;
  ```

- Create the users:

  ```
  CREATE USER admin_user   WITH PASSWORD 'admin_password';
  CREATE USER regular_user WITH PASSWORD 'regular_password';
  CREATE USER backup_user  WITH PASSWORD 'backup_password';
  ```

- Grant privileges:

  ```
  -- Admin user
  GRANT ALL PRIVILEGES ON DATABASE your_database_name
    TO admin_user;
  \c your_database_name
  GRANT ALL PRIVILEGES ON ALL TABLES IN SCHEMA public
    TO admin_user;
  ```

```
...
ALTER DEFAULT PRIVILEGES IN SCHEMA public
  GRANT ALL PRIVILEGES ON FUNCTIONS TO admin_user;

-- Regular user
GRANT CONNECT ON DATABASE your_database_name
  TO regular_user;
...

-- Backup user
GRANT CONNECT ON DATABASE your_database_name
  TO backup_user;
...
```

Replace your_database_name, admin_user, admin_password, regular_user, regular_password, backup_user, and backup_password with your desired names and passwords.

There is little to criticize in the answer, which is somewhat abbreviated here for reasons of space. The difficulty of this question wasn't high. The primary purpose of an AI tool here is to save time. This works even more efficiently if you enter the desired names for the database and the user accounts.

> **Prompt:** Perfect. Could you please repeat the SQL commands (without the explanation) using the following names:
>
> database: bikerental
> admin user: bike_admin
> regular user: bike_user
> backup user: bike_backup

7.3.2 SQL Server: Automated Backup

The classic administration task is to run a backup. Depending on the DBMS and the database size, there are countless variants. In this example, we want to set up an incremental backup process.

> **Prompt:** SQL Server: I would like to set up an automated backup system for my accounting database: a full backup every Sunday night, incremental backups the rest of the week. I want the system to keep the last two weeks of backups before overwriting them. The backups should be stored in a local directory, from where they are synchronized to an external server (not relevant here).

As ChatGPT's answer would be three and a half pages long in this book, we've only quoted the key points here.

To set up an automated backup system for your SQL Server database with the specified requirements, you can use SQL Server Agent for scheduling the backup jobs. Here's a step-by-step guide:

Create a script for the full backup, saving it as FullBackup.sql:

```
DECLARE @BackupFileName NVARCHAR(500)
DECLARE @BackupPath NVARCHAR(500) = 'C:\SQLBackups\'
DECLARE @DatabaseName NVARCHAR(500) = 'YourDatabaseName'
-- Generate file name with timestamp
SET @BackupFileName = @BackupPath + @DatabaseName +
  '_Full_' + CONVERT(NVARCHAR(500), GETDATE(), 112) +
  '.bak'
-- Perform full backup
BACKUP DATABASE @DatabaseName
TO DISK = @BackupFileName
WITH INIT
```

Create a script for the differential backup, saving it as DifferentialBackup.sql:

```
...
-- Perform differential backup
BACKUP DATABASE @DatabaseName
TO DISK = @BackupFileName
WITH DIFFERENTIAL, INIT
```

Create a script for cleaning up old backups, saving it as CleanupBackups.sql:

```
DECLARE @BackupPath NVARCHAR(500) = 'C:\SQLBackups\'
DECLARE @DaysToKeep INT = 14
-- Delete backup files older than @DaysToKeep days
EXECUTE master.dbo.xp_delete_file 0, @BackupPath,
        'bak', @DaysToKeep, 1
```

Create SQL Server Agent jobs: Open SQL Server Management Studio (SSMS), connect to your SQL Server instance, expand the SQL Server Agent node, right-click on Jobs, and select New Job . . .

The solution sounds reasonable, and you can implement the backup system. You'll want to take a look at the original documentation:

Prompt: Can you provide manual links for BACKUP and RESTORE DATABASE?

Certainly! Here are the links to the official Microsoft documentation for the BACKUP and RESTORE DATABASE commands in SQL Server:

- *https://docs.microsoft.com/en-us/sql/t-sql/statements/backup-transact-sql?view= sql-server-ver15*

- *https://docs.microsoft.com/en-us/sql/t-sql/statements/restore-statements-trans-act-sql?view=sql-server-ver15*

The links refer to SQL Server 15 (2019). SQL Server 16 (2022) was the current version at the time we tested this example. However, no major syntax changes are to be expected for the backup and restore commands; in this respect, the links are OK.

The most important rule when setting up a backup system is to test the recovery too!

Prompt: Suppose my main server goes down on a Wednesday. I pull the backups from my off-site backup source. How do I recover my database?

To recover your database from backups when the main server goes down, follow these steps [more details]. Here's a summary of all the commands you might need:

```
-- Restore full backup
RESTORE DATABASE YourDatabaseName
FROM DISK = 'C:\SQLBackups\Accounting_Full_20240602.bak'
WITH NORECOVERY;
-- Restore differential backup from June 3, 2024
RESTORE DATABASE Accounting
FROM DISK = 'C:\SQLBackups\Accounting_Diff_20240603.bak'
WITH NORECOVERY;
-- Restore differential backup from June 4, 2024
RESTORE DATABASE Accounting
FROM DISK = 'C:\SQLBackups\Accounting_Diff_20240604.bak'
WITH NORECOVERY;
-- Restore differential backup from June 5, 2024
RESTORE DATABASE Accounting
FROM DISK = 'C:\SQLBackups\Accounting_Diff_20240605.bak'
WITH RECOVERY;
```

7.3.3 Determining the Schema of a Database

For ChatGPT to help you with questions about an existing database, you must first make the database schema known. The easiest way to do this is to have an SQL file with all the required CREATE TABLE commands. With ChatGPT, you can upload the file using drag and drop. In VS Code or another editor, you can load the file and refer to this file in your question.

But where do you get such a file from? Almost all DBMSs have commands to generate corresponding schema files in text form. The easiest way to do this is to ask ChatGPT about the program you're using.

Prompt: MySQL: How can I save my database schema in a SQL file?

To save your MySQL database schema to a SQL file, you can use the `mysqldump` command with the `--no-data` option. This will create a SQL file containing only the database structure (tables, indexes, etc.) without any of the data.

```
mysqldump -u [username] -p --no-data [dbname] > schema.sql
```

Here's a breakdown of the command [more details]

Prompt: Is there a similar command for SQL Server?

Yes, in SQL Server, you can use the SQL Server Management Studio (SSMS) or a script to generate the schema-only backup. Here's how you can do it using both methods:

Using SQL Server Management Studio (SSMS)

- Open SSMS and connect to your SQL Server instance.
- Right-click the database you want to script.
- Select **Tasks • Generate Scripts**.
- In the wizard, click **Next** to go past the introduction screen.
- Select the specific database objects you want to script (or select the entire database). Click **Next**.
- On the **Set Scripting Options** page, click **Advanced**.
- [more details]

Using a Script with SQLCMD

Alternatively, you can use the `sqlcmd` utility along with a custom query to generate the schema-only script. Here's an example script:

```
sqlcmd -S [server_name] -U [username] -P [password] `
      -d [database_name] `
      -Q "EXEC sp_generate_inserts @schema_only = 1" `
      -o "[output_file].sql"
```

Using PowerShell and SQL Server Management Objects (SMO)

You can also use a PowerShell script to generate the schema. [more details]

With MySQL, the `mysqldump` command is the easiest way. For SQL Server, ChatGPT provides a choice of three solutions, the first of which relates to *SQL Server Management Studio* (SSMS). However, such instructions have the disadvantage that they can often only be reproduced in an identical way if you use a specific version of this user interface.

> **Prompt:** To which version of SSMS does your answer apply?

> The steps provided for generating a schema-only script using SQL Server Management Studio (SSMS) apply to SSMS versions 2012 and later, including the latest versions up to SSMS 19. SSMS has had a consistent interface for these scripting tasks across these versions, so the steps should be applicable.

However, when we ran the preceding prompt in May 2024, version 20.1 was already available, while ChatGPT's training material wasn't that up to date. But with a bit of luck, the instructions will work in the same or a similar way with the latest version.

7.3.4   MariaDB Tuning

A few months ago, we set up a LAMP server (Linux + Apache + MariaDB + PHP). MariaDB has a major impact on the overall performance of the web application running there. Accordingly, we've put a lot of effort into tuning the database server. Would ChatGPT have gotten us there?

> **Prompt:** I have just installed a new LAMP server running Ubuntu 24.04. The server has 64 GB of RAM. It will run as a web server (Apache), mail server (Postfix), and database server (MariaDB 10.11). I need help tuning MariaDB.
>
> MariaDB may use up to 50% of the RAM. There are many databases (about 200) with about 20 tables each. Read performance is critical. (The ratio between SELECT and INSERT/UPDATE/DELETE is about 100:1.)
>
> Please give me five concise recommendations for MariaDB or Linux settings to achieve good performance.

> To optimize MariaDB for your LAMP server setup, follow these five recommendations:
>
> 1. **Adjust the InnoDB Buffer Pool Size** [more details]
>
>    ```
>    [mysqld]
>    innodb_buffer_pool_size = 32G
>    ```
>
> 2. **Optimize Query Cache** [more details]
>
>    ```
>    [mysqld]
>    query_cache_type = 1
>    query_cache_size = 1G
>    query_cache_limit = 1M
>    ```
>
> 3. **Increase the Number of InnoDB Buffer Pool Instances** [more details]
>
>    ```
>    [mysqld]
>    innodb_buffer_pool_instances = 16
>    ```

4. **Tune InnoDB Log File Size and Flush Method** [more details]

```
[mysqld]
innodb_log_file_size = 2G
innodb_flush_method = O_DIRECT
```

5. **Optimize File System and I/O Scheduler:** Use the XFS or ext4 file system with the `noatime` option to reduce disk I/O and set the I/O scheduler to deadline for better performance. [more details]

```
sudo mount -o remount,noatime /
sudo echo deadline > /sys/block/sda/queue/scheduler
```

We weren't entirely convinced by the proposals. Only the first measure is truly uncontroversial.

We also enabled the query cache, independent of ChatGPT's recommendation. However, in recent years, the prevailing opinion has been that the cache positively impacts overall performance only in exceptional cases. A look at the cache statistics suggests that our server is indeed one of these exceptions. Be that as it may, in current MySQL versions, the query cache is no longer available; in MariaDB, it's disabled by default. ChatGPT's recommendation for the query cache likely results from countless old websites that were used in training the language model. This is quite a common problem; ChatGPT tends to prefer outdated measures because they appear more frequently in the training material than modern guidelines.

Measures 3, 4, and 5 are unlikely to result in any measurable differences in performance. If you really want to use the deadline scheduler, note that it's called `mq-deadline` in current kernel versions (including Ubuntu 24.04, which was mentioned in the prompt).

However, we're missing another measure in the list of suggestions, which would actually be obvious with the setup described in the prompt: increasing the limit for simultaneously open files (`mysql soft nofile 65535` and `mysql hard nofile 65535` in /etc/security/limits.conf).

In short, as a tuning aid, ChatGPT can provide food for thought, but the quality of the response is nowhere near as good as when compiling SQL commands.

7.4   Client Programming

Database design and DBMS administration usually take far less time than developing the client code, be it for a web interface, for an app, or for business classes that are located between the DBMS and the actual client code. Naturally, AI tools also help with coding in this case.

In our experience, an AI chat is only useful at the design stage, that is, when you select a suitable database library or develop and test initial functions. As a result, it's much more efficient to edit the code for detailed problems directly using an AI-supported editor. All you need to do is use a function name like getRecentOrders, and GitHub Copilot or another AI tool will suggest suitable code. The more comparable functions or methods are available in the code base as a guide, the better this will work.

We'll refrain from printing longer code examples in this section, but merely provide a few suggestions for prompts. A basic prerequisite for meaningful answers is that the AI tool has access to the database schema. However, we've found that many AI tools are unable to extract column names correctly even with the schema. Sometimes, it helps to specify only the relevant tables in the prompt in the form of CREATE-TABLE commands.

> **Prompt:** PHP/MySQL/MySQLi: I need an alphabetical list of all customers who have placed an order in the past six months. The result should be displayed in a six-column HTML table. Each result is a link to a detail page customer?id=xxx.

> **Prompt:** Node.js/MySQL: I want to write a function that returns the customer list (as above) via a REST request (GET customers/lastOrderSince=yyyyy-mm-dd).

> **Prompt:** Python/PyMySQL: I would like to send a promotional offer to all customers who haven't placed an order in more than six months. Can you provide a script?

> **Prompt:** PHP/MySQLi: How can I bundle multiple SQL commands into one transaction?

> **Prompt:** C#: Which library should I use to access an SQLite database?

> **Prompt:** Bash: I have a list of MariaDB databases in mydbs.txt. I need to do a schema update on all databases. The necessary ALTER TABLE commands are in v32.sql. Can you provide a script?

Chapter 8
Scripting and System Administration

There are few areas of coding where AI support—used correctly!—works as well as when you use Python or Bash for scripting. In this context, the nature of scripts plays a crucial role. Often, relatively few lines of code are sufficient to complete a clearly defined task. While the code for large programming projects is spread across dozens, often hundreds of files, a script can often be formulated in a single file. Accordingly, there are hardly any dependencies or context issues that overwhelm AI language models.

We had actually expected to have similarly good experiences with scripting on Windows using PowerShell. But the result was rather disappointing. The GPT-4o model used in our tests delivered (as of mid-2024) much poorer quality code with significantly more errors. Of course, there is less training material for PowerShell available on the internet than for Python, but it was surprising that the difference would be so obvious.

Nevertheless, in this chapter, we present a whole range of examples where ChatGPT, GitHub Copilot, and other AI tools have helped us. Once again, our aim is to present you with a wide range of possible applications. Perhaps there is an idea or two that you haven't thought of yet?

In this chapter, we also go a little beyond the scripting horizon and look at secondary topics such as regular expressions and system administration.

System administration isn't really a classic coding topic and is therefore outside the focus of this book. But in real life, many developers and programmers are also responsible for setting up Docker containers, Git repositories, database systems, and servers. ChatGPT and other AI tools are perfectly capable of helping with this.

8.1 Scripting

From our point of view, the ideal working environment for writing new scripts is an editor with AI support (e.g., Visual Studio Code [VS Code] plus GitHub Copilot) and a dedicated chat window—whether as part of the editor or separate from it in a web browser. The AI chat provides help with the basic design of parts of the script and with researching the commands, cmdlets, or modules used. The editor supports code input, documentation, and debugging. The following pages present a colorful mix of examples.

8.1.1   Bash: Loop Over a Text File

You want to process a text file in a Bash script and skip the first line with the column label. What is the easiest way? You ask an AI tool!

> **Prompt:** Bash: I want to loop over all lines of file but skip the first line.

> In Bash, if you want to loop over all lines of a file but skip the first line, you can use the following script:
>
> ```
> # Assuming 'file.txt' is your file
> tail -n +2 file.txt | while IFS= read -r line
> do
>   echo "$line"
> done
> ```
>
> Here's a breakdown of the script [more details]

ChatGPT provides a solution that is as short as it is ideal. The rather lengthy explanations of the script are primarily of interest to Bash beginners.

8.1.2   PowerShell: Sorting Photos

You want to use a PowerShell script to analyze the exchangeable image file format (EXIF) data of an entire collection of photos. The image files should be moved accordingly to directories named yyyy-mm, such as 2024-06 for photos shot in June 2024.

> **Prompt:** PowerShell script: I have a lot of JPEG files to organize. I want to read the EXIF date and then move the files to folders like yyyy-mm (year/month). Files without EXIF data should be moved to a special folder. Can you help me?

> Sure! Here is a PowerShell script that reads the EXIF date from JPEG files and organizes them into folders based on the year and month. Files without EXIF data will be moved to a separate folder.
>
> Script and explanation . . .

ChatGPT provides an approximately 60-line script that uses the System.Drawing system library to analyze the EXIF data. Although this solution is relatively cumbersome, it has one major advantage: no additional tools or libraries need to be installed.

However, an initial test run fails: the script opens each image file for EXIF analysis, but doesn't close the file again. This is why Move-Item fails with an error message for every file.

The original script still has one flaw: it only takes *.jpg files into account, but not *.jpeg files. For this reason, you need to ask ChatGPT for rectification.

> **Prompt:** Move-Item causes an error: "The process cannot access the file because it is being used by another process." I guess the script needs to close the file before it can move it. Please also process *.jpeg files.

> You're correct. The image files should be closed before moving them to avoid the "file in use" error. Additionally, we'll include processing for *.jpeg files as well. Here's the updated script:
>
> ```
> ...
> $jpegFiles = Get-ChildItem -Path $sourceDir `
>                       -Filter *.jpg, *.jpeg
> ...
> ```

ChatGPT recognizes its mistake and corrects the script, but immediately inserts a new error: the -Filter option expects a single character string as an argument, not an enumeration.

> **Prompt:** $jpegFiles = Get-ChildItem -Path $sourceDir -Filter *.jpg, *.jpeg causes an error:
>
> Cannot convert System.Object[] to the type System.String required by parameter Filter. Specified method is not supported.

> I apologize for the mistake. The -Filter parameter only accepts a single string. We can handle both .jpg and .jpeg files by using Get-ChildItem twice or by using -Include with -Recurse. Here is the corrected script:
>
> ```
> ...
> $jpegFiles = Get-ChildItem -Path $sourceDir `
>                       -Include *.jpg, *.jpeg -Recurse
> ...
> ```

Only the script delivered in the third attempt works satisfactorily. The -Recurse option now used has the additional benefit that PowerShell processes not only the image directory but also all subdirectories. This behavior should actually be self-evident, but was admittedly not required in the original prompt.

What is confusing about this example is the triviality of the errors. While the basic concept of the script is fine on the first attempt, the errors it contains are to a certain extent beginner's mistakes. They undermine trust in the AI tool.

8.1.3   Python: SQLite Application

You want to organize internal data with SQLite in a Python script. You know SQLite as a minimalist library for small database applications, but you've never used SQLite together with Python.

Prompt: I am working on a Python script and want to use SQLite as a local database. Which module should I use? Do I need to install any additional libraries? Can you provide a short Hello-World-sample?

To use SQLite in Python, you can utilize the built-in sqlite3 module. There's no need to install any additional libraries since sqlite3 is included with Python's standard library. This module provides a lightweight disk-based database that doesn't require a separate server process and allows access using a nonstandard variant of SQL.

Here's a short "Hello World" example of how to use SQLite with Python:

```
import sqlite3
conn = sqlite3.connect(':memory:')
cursor = conn.cursor()
cursor.execute("""CREATE TABLE hello_world
                (id INTEGER PRIMARY KEY, message TEXT)""")
cursor.execute("""INSERT INTO hello_world (message)
                VALUES ('Hello, World!')""")
conn.commit()
cursor.execute('SELECT * FROM hello_world')
print(cursor.fetchall())
# Output: [(1, 'Hello, World!')]
conn.close()
```

This script performs the following steps [more details].

You can modify this example to interact with a file-based database by replacing ':memory:' with a file path, like 'example.db'.

ChatGPT provides exactly the right starting point for your own experiments. You can perform a short test run to make sure that the script really does work. Then, try to adapt the script to your needs via AI chat.

We further assume that you use VS Code and GitHub Copilot. Of course, you can use any other comparable combination of editor and code wizard.

> **Prompt:** I want to store a URL, an IP address (IPv4/IPv6) and a timestamp in a table. Can you adapt `hello-world.py` accordingly?

GitHub Copilot then generates new code and displays it in the chat section (not printed here). However, you must manually transfer the code into your script.

An alternative procedure is to select the relevant code section and formulate your code request via `Ctrl`+`I` or `cmd`+`I` in an inline chat. This has the advantage that VS Code now actually suggests changing your code file. You can accept or reject the new code using **Accept** or **Discard**.

> **Prompt:** Please save the database persistent in `log.db`. If this file exists, open it. If not, create a new database with the `web_data` table.

However, the inline chat process isn't really "smart" yet. First, the logic of the code is clearly wrong here.

Figure 8.1 GitHub Copilot Didn't Do a Great Job: Faulty New Code Logic and Incomplete Integration into the Existing Code

The test as to whether the database file exists must be carried out *before* the database is opened or created. Second, the new code only partially matches the replaced code. Imports are missing, parts of the code are redundant, and so on. If you accept the AI suggestion, you'll need to edit the code extensively afterwards.

A Look into the Future

This example explains why we prefer a chat function that is separate from the code when making major code changes and why we prefer to carry out the actual code conversion manually. But, of course, it's only a matter of time before code wizards are smart enough to make complex code changes on their own.

8.1.4   Bash Troubleshooting

ChatGPT can also help with troubleshooting. Some time ago, we wanted to process a comma-separated values (CSV) file. The two columns for URL and database were separated by semicolons. We've tried to split the columns using cut -d;, but this fails:

Prompt: The following Bash script doesn't work. Why?

```
tail -n +2 unused.csv | while IFS= read -r line; do
    url=$(echo $line | cut -d; -f1)
    db=$(echo $line | cut -d; -f2)
    echo "url=$url db=$db"
done
```

ChatGPT version 4 and higher recognizes the error immediately. (What about you?)

The issue with your script is likely due to the incorrect handling of the delimiter in the cut command. The delimiter needs to be specified properly, and special characters (like the semicolon ; in your case) should be quoted to ensure they are interpreted correctly by the shell. Here is how you can fix the code:

```
tail -n +2 unused.csv | while IFS= read -r line; do
    url=$(echo "$line" | cut -d';' -f1)
    db=$(echo "$line" | cut -d';' -f2)
    echo "url=$url db=$db"
done
```

Older GPT versions failed with this example, as did the local models llama3:8b and gemma2:9b.

8.1.5 Python: Recursive Search for Files

The starting point for this example was the work on a script to recursively process all files with predefined identifiers starting with a start directory (see also Chapter 10, Section 10.4).

Prompt: Python: I need a function that returns all files in a directory (recursively) with given extensions. I will pass two parameters, directory path and a list of extensions (e.g., [".js", ".php"]). The code should run on Linux, macOS, and Linux. Could you please help?

Certainly! Below is a Python function that will search through a directory recursively and return a list of all files that match the given extensions. This function uses the os and os.path modules for directory traversal.

```
def get_files_with_extensions(directory, extensions):
    matching_files = []
    for root, _, files in os.walk(directory):
        for file in files:
            if any(file.endswith(ext) for ext in extensions):
                matching_files.append(os.path.join(root, file))
    return matching_files

# example usage
print(get_files_with_extensions("/home/kofler",
                                [".js", ".php"])
```

[more details]

8.1.6 Other Prompt Examples

In this section, we've collected some more prompts from our real-life work, but have spared ourselves the presentation of the results or the discussion of the results. The prompts should simply serve as a suggestion.

Prompt: PowerShell: I need a script to scan my download directory for files older than 30 days. The script lists the files alphabetically (including size information) and offers to delete them after explicit confirmation (input of "yes").

Prompt: Bash: I want to write a script that symmetrically encrypts all files that are passed as parameters. The encryption key (binary) is the content of /etc/secretkey. Add .crypt to the new file name, and save it in the same directory as the old one. With the --delete-original option, delete the original files.

Prompt: Bash: Write a script that uses ImageMagick to create JPEG thumbnails of the front page of all PDF files in a directory. For the thumbnail file names, replace the .pdf extension with .jpg. Skip PDF files that already have a thumbnail.

Prompt: Python: I have a CSV file like this:

[20 lines of CSV text]

I need a script to visualize the column "energy consumption" over time. Can you help?

Prompt: Python/matplotlib: How can I label only every second point on the x-axis?

Prompt: Python: I want to display an image and interactively mark points on it. I need the x and y coordinates of these points. What are my options?

8.2   Example: Converting a Python Script to PHP Code

The following example also comes from our everyday work, but is somewhat more extensive. The starting point is a Python script about 100 lines long. It reads a text file containing database names. The script then analyzes more than 100 similar databases and creates a statistical summary of customer activity. The result is a hard-to-read ASCII table that is sent via email every month.

We want to turn it into a PHP page that can be called at any time. The result is supposed to be displayed as an HTML table. For the code translation, we used ChatGPT 4o; all other work was done in VS Code with GitHub Copilot.

In this example, we'll focus on explaining the prompts. We've refrained from printing the very project-specific code in PHP or Python as it doesn't really contribute to understanding the procedure.

Prompt: I have a Python script that is run monthly to summarize account activity. I want to convert this script to PHP. The result is to be presented in an HTML table instead of ASCII. Can you help please?

[100 lines of Python code]

Certainly! Here's how you can convert the provided Python script to PHP and modify it to present the result in an HTML table.

[170 lines of PHP code]

[more details, explanation, running the script]

The resulting code leaves a good impression at first glance. But we have two fundamental problems:

- First, ChatGPT mixes HTML code with embedded PHP code when outputting the results. This is quite typical of PHP, but we prefer consistent PHP code that performs all output via echo. This makes it easier for us to maintain an overview of the code.

- Secondly, the code isn't organized as a function. However, we want to separate the output of statistical data from other subfunctions.

So, we ask for improvements. For this purpose, we first select a code section. The following prompt refers to the selected code. result is a variable defined in the previous code that contains the data previously determined from the databases in an array. The highlighted code analyzes this data and turns it into an HTML table.

Prompt: Could you convert this code into a function (with parameter $result) using echo to output?

Certainly! Here's how you can convert the provided Python script to PHP and modify it to present the result in an HTML table.

[entire code, including the new function displayAccountActivitySummary ...]

On this basis, we copy the code generated by ChatGPT to VS Code and make various adjustments there. In particular, we use a class for establishing the connection to the MySQL database that is already available for this purpose in our PHP project. In VS Code, we're supported in our work by GitHub Copilot.

Just fifteen minutes after starting work, we're able to carry out an initial test, which is generally satisfactory. For the next fifteen minutes, we optimize the HTML table layout, improve the labeling of the columns, and add a legend to the table. GitHub Copilot surprises us with almost clairvoyant suggestions for the texts, which we adopt with minimal changes.

ChatGPT has adopted the criteria for sorting the results unchanged from the Python script. This basically works, but we want to have a second sorting criterion. However, we have little experience with the PHP function usort used by ChatGPT.

Prompt: PHP/usort: How do I sort using two criteria instead of one?

We had actually expected a universal answer. Instead, ChatGPT responds with a few lines of code that uses exactly the attribute of the results we had in mind anyway as the second sorting criterion. Without any further changes, the code does what we intended it to do but didn't say. That's a bit uncanny.

8.2.1 Conclusion

The path to a new PHP script presented here is a prime example of how well AI tools can work. We were surprised ourselves at how quickly we reached our destination. Before you object that this is a contrived example, note that we carried out the work outlined here for a project that has nothing to do with this book. We only came up with the idea of presenting the prompts in this chapter at a later stage. So, this very example originates precisely from our real-life work.

8.3   Regular Expressions

Regular expressions are character strings that describe a search pattern for text. In this context, specific rules must be followed that aren't very intuitive. Various Linux commands such as grep and sed, as well as almost all programming languages, support regular expressions. You can use them to filter out specific information from a text, validate input, or replace recognized patterns with new texts.

AI tools can now help you with what used to be a laborious compilation of regular expressions in the past. Please note, however, that there are several regular expression syntax variants, including *Portable Operating System Interface* (POSIX), *POSIX Extended Regular Expressions* (ERE), and *Perl Compatible Regular Expressions* (PCRE). It's therefore important that you always specify which tool or programming language you use. ChatGPT and others can then also contribute the code for applying the regular expression.

> **Prompt:** Please give me a regular expression for email addresses (PCRE syntax).

Sure, here's a regular expression for matching email addresses:

`[A-Za-z0-9._%+-]+@[A-Za-z0-9.-]+\.[A-Za-z]{2,}`

- `[A-Za-z0-9._%+-]+`: Matches one or more of the characters within the square brackets, which include letters (both uppercase and lowercase), digits, and certain special characters commonly found in email addresses like period (.), underscore (_), percent (%), plus (+), and hyphen (-).
- `@`: Matches the @ symbol.
- `[A-Za-z0-9.-]+`: Matches one or more letters (both uppercase and lowercase), digits, and certain special characters commonly found in domain names like period (.) and hyphen (-).
- `\.`: Matches a period (dot), which is escaped with a backslash because a period is a special character in regular expressions.
- `[A-Za-z]{2,}`: Matches two or more letters (both uppercase and lowercase) for the top-level domain (TLD). This ensures that the TLD portion of the email address consists of at least two letters.

This regex pattern should cover most valid email addresses, but it's worth noting that email address validation with regular expression can be quite complex due to the various standards and possible edge cases.

It can also be shorter. However, the simpler expression requested in this second example is less precise.

Prompt: Is a shorter regular expression possible?

Yes, a shorter regex is possible. Here's a more concise version:

`\S+@\S+\.\S+`

- `\S+`: Matches one or more non-whitespace characters. This will match any sequence of characters before and after the "@" symbol.
- `@`: Matches the "@" symbol.
- `\S+`: Matches one or more non-whitespace characters. This will match any sequence of characters in the domain part of the email address.
- `\.`: Matches a period (dot), which is escaped with a backslash.
- `\S+`: Matches one or more non-whitespace characters. This will match any sequence of characters in the top-level domain (TLD) part of the email address.

This shorter regex pattern also captures most valid email addresses, but like the longer version, it may not cover all possible edge cases or conform exactly to all email address standards.

8.3.1   Searching and Replacing Dates

For this example, we were looking for a Python script that replaces all dates in a text file that are written as MM/DD/YYYYY and DD.MM.YYYY with ISO data, that is, YYYY-MM-DD.

Prompt: Python: I want to replace all international dates (12/31/2024) and German dates (31.12.2024) by ISO dates (2024-12-31) within a text file.

Here is a Python script that reads a text file, replaces all international and German date formats with ISO date format, and writes the changes back to the file:

```python
import re

def convert_dates_to_iso(file_path):
    with open(file_path, 'r') as file:
        content = file.read()

    # Pattern for international dates (MM/DD/YYYY)
    content = re.sub(r'(\b\d{1,2})/(\d{1,2})/(\d{4}\b)',
                     r'\3-\1-\2', content)
    # Pattern for German dates (DD.MM.YYYY)
```

```
content = re.sub(r'(\b\d{1,2})\.(\d{1,2})\.(\d{4}\b)',
                 r'\3-\2-\1', content)

with open(file_path, 'w') as file:
    file.write(content)

# Example usage
convert_dates_to_iso('path_to_your_file.txt')
```

What is remarkable about the solution is the handling of groups. Groups can be defined within a regular expression using parentheses. In the second replacement expression, \1 denotes the content of the first group, \2 that of the second group, and so on. In this way, the new 2024-12-31 string can be created from 31.12.2024.

8.3.2   File Search Using grep

The starting point for the following example was an HTML file and a directory with image files. On the one hand, we wanted to delete all image files that weren't linked in the HTML file and, on the other hand, ensure that the directory with the image files is complete, that is, that the HTML file doesn't reference an image in this directory that doesn't even exist there.

Prompt: Regular expression/grep: I want to extract all image file names from an HTML file. The image file names always start with "screenshots/" and end with ".jpg" or ".png". Can you help?

You can use a regular expression with grep to extract image file names from an HTML file. Here's a command you can use:

```
grep -o 'screenshots/[^"]*\.jpg\|screenshots/[^"]*\.png' \
  your_html_file.html
```

Explanation:

- -o option: Print only the matched parts of a matching line.
- 'screenshots/[^"]*\.jpg\|screenshots/[^"]*\.png': This regular expression matches strings that start with screenshots/, followed by any characters except ", and end with either .jpg or .png.

In the follow-up question, we ask ChatGPT for a way to delete the unnecessary images. The result is a Bash script.

Prompt: Great. Now I want to delete all image files that are *not* in the Grep result. (Directory screenshots contains more images than necessary.)

You can achieve this by using a combination of `grep` and `rm` commands in a shell script. Here's a script that should do the job:

```bash
#!/bin/bash
# the directory containing the image files
image_dir="screenshots"
# the HTML file
html_file="your_html_file.html"

# use grep to find all image file names
image_files=$(grep -o \
  'screenshots/[^"]*\.jpg\|screenshots/[^"]*\.png' \
  "$html_file")

# iterate over all files in the image directory
for file in "$image_dir"/*; do
    # Extract the filename from the full path
    filename=$(basename "$file")
    # check if the filename is not in the list of image files
    if ! grep -q "screenshots/$filename" <<< "$image_files";
    then
        # if not found, delete the file
        rm "$file"
        echo "Deleted $file"
    fi
done
```

Naturally, you'll only try such a script after you've made sure that there is a backup of all files! Fortunately, everything works perfectly here.

Prompt: Perfect. Now we need to do it the other way round. I need another script which lists all image files referenced in the HTML file that are missing in the screenshots directory.

```bash
#!/bin/bash
image_dir="screenshots"
html_file="your_html_file.html"
referenced_files=$(grep -o \
  'screenshots/[^"]*\.jpg\|screenshots/[^"]*\.png' \
  "$html_file")

# iterate over all files referenced in the HTML file
for ref_file in $referenced_files; do
```

```
    # extract the filename from the path
    filename=$(basename "$ref_file")
    # check if the file exists in the image directory
    if [ ! -f "$image_dir/$filename" ]; then
        # if not found, print the missing file
        echo "Missing file: $filename"
    fi
done
```

8.4   System Administration

As a developer, you're of course primarily responsible for coding. In real life, however, the clear distinction between administration and software development sometimes becomes blurred. You need to set up Docker containers, network setups, virtual machines, or database servers (even if only as a test environment). As the following examples show, AI tools can support you with clear instructions.

We didn't want to call this section "Linux System Administration," but we have to admit that the following pages primarily deal with Linux. This doesn't (only) have to do with personal preference, but mainly with the fact that Linux system administration is done in text mode. You run commands in the terminal, change configuration files, and so on. A language model can be a good help here.

In contrast, Windows administration is often carried out in graphical user interfaces. This doesn't suit the nature of language models. In addition, as of early 2025, ChatGPT and others aren't yet able to support explanations with screenshots of dialogs or setup wizards.

8.4.1   SSH Authentication by Key

The first example is about automatically copying files to a Linux server via scp. Secure Shell (SSH) keys must be used to avoid having to enter a password each time a script is executed.

> **Prompt:** I want to use scp to back up files to a Linux server. The script should run without interaction. What's a safe way to accomplish this?

To use scp for backing up files to a Linux server without requiring interaction, you can use key-based authentication. Here's a step-by-step guide to set this up.

- First, generate an SSH key pair on your local machine:

    ```
    ssh-keygen -t rsa -b 4096 -C "your_email@example.com"
    ```

Follow the prompts. When asked for a passphrase, you can choose to leave it empty for automated processes, but it's recommended to use a passphrase for added security. The key pair is typically saved in `~/.ssh/id_rsa` and `~/.ssh/id_rsa.pub`.

- Use `ssh-copy-id` to copy your public key to the remote server:

```
ssh-copy-id username@remote_server_ip
```

This command will prompt for your password on the remote server. After successful execution, your public key will be added to the `~/.ssh/authorized_keys` file on the remote server.

- Verify that you can connect to the remote server without being prompted for a password:

```
ssh username@remote_server_ip
```

ChatGPT then provides an example of a simple backup script and its automation via Cron. We've deleted this part of the output here.

Basically, SSH key authentication also works if the script is supposed to run on Windows. However, the setup is a little more complicated. ChatGPT explains how to manually set up the public part of the key on the server.

> **Prompt:** My backup script should run on Windows. However, `ssh-copy-id` isn't available. Can I work around this problem?

Yes, you can work around this issue on Windows by manually copying your SSH public key to the remote server. First, generate an SSH key pair using the built-in OpenSSH client in Windows 10/11.

```
ssh-keygen -t rsa -b 4096 -C "your_email@example.com"
```

Follow the prompts to save the key (by default, it will be saved to `C:\Users\your_username\.ssh\id_rsa` and `C:\Users\your_username\.ssh\id_rsa.pub`).

Since `ssh-copy-id` isn't available on Windows, you need to manually copy the contents of the public key file (`id_rsa.pub`) to the `~/.ssh/authorized_keys` file on the remote server. Open the `id_rsa.pub` file in a text editor and copy its contents. Connect to your remote server using an SSH client. Once logged in, open or create the `~/.ssh/authorized_keys` file on the remote server:

```
nano ~/.ssh/authorized_keys
```

Paste the copied public key into the `authorized_keys` file and save it.

8.4.2   Docker Setup

To test a database client, you need a local MySQL server. You don't want to perform a "real" installation, but want to use Docker. How does that work?

Prompt: I have a local program that needs access to a MySQL database. How can I set up a Docker container accordingly?

ChatGPT provides a detailed seven-point description, starting with the Docker installation and ending with a Python application example. We've only removed the three central Docker commands here:

```
docker network create my_network
docker pull mysql:latest
docker run --name my_mysql_container --network my_network \
  -e MYSQL_ROOT_PASSWORD=rootpassword \
  -e MYSQL_DATABASE=mydatabase \
  -e MYSQL_USER=myuser \
  -e MYSQL_PASSWORD=mypassword \
  -p 3306:3306 -d mysql:latest
```

Basically, it works. Creating a separate network for the container is completely superfluous here, but doesn't interfere. In the last point, ChatGPT points out that all data stored in the database will be lost as soon as the container terminates. To prevent this, you must create a *volume* when creating the container. You can reuse this volume later and transfer the database from one container to the next.

If you want to persist data even after the container stops, you can mount a volume. Modify the Docker run command to include a volume:

```
docker run ... -v my_mysql_data:/var/lib/mysql
```

ChatGPT will tell you how to transfer the data to a new container if required in a follow-up question. We've only quoted the central sentence of the answer here.

Prompt: After ending the first container, how would I create a new container reusing the volume?

When creating the new container, use the same volume (my_mysql_data) ...

8.4.3   Proxy Server

Modern web applications and REST APIs are usually connected to the internet via a proxy server. AI tools can explain how they work and how to set them up:

Prompt: I want to build a REST application using Python and Django. The application will run in a Docker container. Do I need a proxy server in this case to make my application visible to the internet?

ChatGPT first explains how a reverse proxy server works and its advantages and then provides instructions for a corresponding Docker setup. If you do something like this for the first time, you'll probably only get there with a few additional questions. However, if you already have some experience in this regard, the Dockerfile and docker-compose.yml files compiled by ChatGPT are an excellent starting point.

In the context of deploying a REST application using Python and Django in a Docker container, whether you need a proxy server depends on your deployment architecture and requirements. Here's a breakdown to help you understand when and why you might need a proxy server. [more details]

Here's a high-level overview of setting up a Django REST application with Docker and Nginx as a reverse proxy:

- Dockerize Your Django Application [more details]
- Create a Dockerfile [more details]
- Create a docker-compose.yml [more details]
- Set Up Nginx as a Reverse Proxy [more details]
- Run Your Application

ChatGPT is currently only able to visualize the data in a rather dull ASCII representation. Here, GPT-4o doesn't yet live up to the "omni" promises.

Prompt: Can you visualize the setup in a diagram? Please use ASCII art.

Certainly! Here's a textual description of the diagram for the setup described:

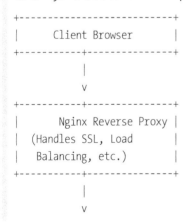

```
+---------------------------+
|      Client Browser       |
+-----------+---------------+
            |
            v
+-----------+---------------+
|      Nginx Reverse Proxy  |
|  (Handles SSL, Load       |
|   Balancing, etc.)        |
+-----------+---------------+
            |
            v
```

```
+-----------+----------------+
| Docker Container Network   |
|                            |
|   +----------+  +---------+|
|   |   Web    |  |   DB    ||
|   | Container|  |Container||
|   +----------+  +---------+|
|    (Django)      (Postgres)|
+----------------------------+
```

8.4.4 Fixing Network Issues in a Virtual Machine with a Web Server

You've installed Linux, a web server, and various other tools in a virtual machine. But you fail to access the web server from your host. Can ChatGPT help?

Remember to be as precise as possible. Which virtualization program? Which Linux distribution? Which web server? Which host operating system? What error message? What have you already tried to narrow down the error?

> **Prompt:** I am using VirtualBox on my Windows notebook. I have set up a virtual machine with Fedora Linux. Within Fedora, I have installed Apache.
>
> - `systemctl` status `httpd` says it is running fine.
> - `hostname -I` reveals that my IP address is 10.0.2.15.
>
> However, on my host system, I can't access the website. Whenever I try to open *http://10.0.2.15*, the web browser displays `ERR_CONNECTION_TIMED_OUT`. What could cause this error?

The issue you're experiencing is likely due to the network configuration of your VirtualBox setup. By default, VirtualBox often uses NAT (Network Address Translation) for networking, which can cause issues with accessing the guest machine from the host. Here are a few steps to resolve this:

- Change Network Adapter to Bridged Adapter [more details]
- Determine New IP Address [more details]
- Update Firewall Rules. Ensure that the firewall on Fedora allows HTTP traffic. First, check the status of the firewall:

  ```
  sudo firewall-cmd --state
  ```

  If the firewall is running, allow HTTP traffic:

  ```
  sudo firewall-cmd --permanent --add-service=http
  sudo firewall-cmd --reload
  ```

- Access the Web Server [more details]

In our experience, ChatGPT's response is perfect. The AI tool points out the two most common causes of errors. First, the default network configuration of VirtualBox is unsuitable. You need to switch to *bridged networking*. Second, a firewall may be active in the virtual machine. (This is definitely the case with Fedora.) You must define an exception rule that allows network traffic for HTTP.

If you find it difficult to follow the instructions, you can ask follow-up questions. In the interplay between host system, virtualization program, and virtual machine, it's not always easy to identify the correct location for the individual configuration steps.

> **Prompt:** Should I change the firewall settings on my Windows notebook or within the virtual machine?

> To ensure that your Fedora Apache web server is accessible from your Windows host, you should primarily focus on the firewall settings within the virtual machine (Fedora). [more details]

8.4.5   Linux Server Setup

You want to set up a Linux server. You use Redundant Array of Independent Disks 1 (RAID-1), that is, two solid-state drives (SSDs) whose contents are mirrored. If one of the two volumes crashes, you don't have to worry about losing any data.

Now the boot files are usually only on the first disk. If that disk fails, you'll not lose any data, but you'll no longer be able to boot your computer. This can be solved by correctly configuring the GRUB boot loader on *both* disks. But how does that work?

> **Prompt:** Linux: I want to set up a server with two disks using RAID-1. The server should be bootable no matter whether the first or the second disk fails. Can you help?

ChatGPT provides a detailed instruction that starts with partitioning the disks, then explains the RAID setup, and finally covers the installation of GRUB. We've only reproduced the crucial point for the question here:

> Ensure GRUB is installed on both disks. You can do this manually after the installation:

```
sudo grub-install /dev/sda
sudo grub-install /dev/sdb
sudo update-grub
```

It's therefore important that the GRUB boot loader is installed on *both* disks, which typically isn't the case. Please note that you must use the device names dev/nvme0n1 and /dev/nvme1n1 instead of /dev/sda and /dev/sdb for modern servers with Non-Volatile Memory Express (NVMe) SSDs.

8.4.6   Additional Prompts

We want to round off this section with a few prompt examples without going into the results in detail.

> **Prompt:** Bash, find command: Is there a way to search for files within a certain date range?

> **Prompt:** PowerShell: How do I list files sorted by modified date?

> **Prompt:** Linux: I want to set up a shared directory for five students to work together. By default, each student should be able to read and write files created by other students in the group. How do I do this?

> **Prompt:** Linux: I have a Logical Volume Management (LVM) setup. I need to increase the size of one of my logical volumes, but the volume group is already fully used. Therefore, I plan to add a second disk. How can I add this disk in my LVM system? How do I increase the size of the logical volume and the file system on it?

> **Prompt:** Linux Server with Postfix mail server: I can send but I can't receive mail. Can you help me debug this?

> **Prompt:** Linux Server with nginx web server. How can I set up a Let's Encrypt certificate for the main domain and three subdomains (*example.com*, *www.example.com*, *help.example.com*, and *support.example.com*).

> **Prompt:** Ubuntu server: I would like to set up a firewall that blocks all incoming traffic with these exceptions: ping, SSH, HTTP, and HTTPS. How should I proceed?

> **Prompt:** ReadyMedia: Can I use ReadyMedia for radio streams (.m3u)? If not, what are my options for a simple setup on the Raspberry Pi?

> **Prompt:** Linux: I want to implement geo-blocking with a nftables firewall to protect my services against attacks from certain countries. What are my options?

8.5   Example: "wget" Script Plus "tmpfs" Configuration

In this section, we present a somewhat longer example from our real-life work that combines script programming and system administration.

A tool used on a Linux server is supposed to turn an HTML file into a PDF document. However, the rather old version of the tool was incompatible with modern HTTPS encryption protocols. For various reasons, updating to a new version of the tool was out of the question.

This gave rise to the following idea: it should be possible to use wget to download the HTML file, including all additional files referenced there (Cascading Style Sheets [CSS], images, etc.) to a temporary directory. The local files could then be converted to PDF format without network problems.

---

wget

wget is a command commonly used on Linux to download a file via FTP or HTTP. wget also has countless special functions that are controlled by options. Among other things, the command can download entire websites. It adapts the links within the HTML pages so that the mirrored website can be viewed locally in the web browser. The trick when using wget is to pass the right combination of options.

8.5.1   Developing a "wget" Script

So, the goal was to develop a Bash script that takes all the parameters intended for the tool, creates a temporary directory, downloads the HTML file passed in the first parameter, calls the HTML-PDF converter, and finally cleans everything up. The script was developed in VS Code, with an SSH connection to the server in the editor.

Prompt: I want to write a small Bash script which I pass a URL (first parameter) and some options. The script creates a temporary directory, downloads the URL plus all referenced files (wget relink), calls another command to process these local files, and finally deletes the temporary directory. Can you help? (Short, concise code preferred.)

Certainly! Below is a concise Bash script to achieve what you described:

```
#!/bin/bash
URL=$1
shift    # shift the arguments to remove the URL
# Create a temporary directory
TMPDIR=$(mktemp -d)
# Ensure the temporary directory is deleted on script exit
trap 'rm -rf "$TMPDIR"' EXIT
# Download the URL and all referenced files into the
# temporary directory
wget --mirror --convert-links --adjust-extension \
     --page-requisites --no-parent -P "$TMPDIR" "$URL"
```

```
# Call another command to process these local files
your_command "$TMPDIR" "$@"
```

Explanation

- **Trap**: Ensures that the temporary directory is deleted when the script exits, regardless of success or failure.
- wget **Command**: Uses wget to download the URL and all referenced files into the temporary directory. The options used are:
  - --mirror: Enables options suitable for mirroring.
  - --convert-links: Converts the links to make them suitable for offline viewing.
  - --adjust-extension: Adds the appropriate extensions to files.
  - --page-requisites: Downloads all necessary files for displaying the page.
  - --no-parent: Prevents wget from following links outside the given URL.
  - -P "$TMPDIR": Specifies the directory to save the files.
- **Processing Command**: Replace your_command with the actual command you want to run on the downloaded files. The script passes the temporary directory and any additional options to this command.

The first version of the script already looks quite promising. We've shortened the explanations of the code so that the example remains clear. To try it out, we've adopted the suggestion nearly unchanged and only replaced your_command with the absolute path to the HTML PDF converter.

During testing, however, we encountered a few issues, some of which we fixed ourselves and others with AI support. The following prompts document this iterative process. We've refrained from reproducing the rather lengthy answers.

> **Prompt:** I need to pass $URL to your_command. However, $URL needs to be modified to refer to the local start point generated by wget.

> **Prompt:** $URL can contain all sort of special characters. Can I get wget to use index.html as a starting point to the local file tree?

ChatGPT doesn't find a suitable option (as far as we know, there is none) and suggests searching for *.html files in the temporary directory by using find and then using the first file that gets found. However, to us this seems prone to errors:

> **Prompt:** Is it safe to simply take the first result? Perhaps there are other *.html files in the directory. Would it make sense to extract the base name from $URL and use it as a search pattern with find?

ChatGPT now provides code that meets our requirements. However, it turns out that wget changes the HTML file too extensively. wget --mirror not only downloads the files required to display the HTML page but also processes *all* links contained in the HTML page. As a result, external links in the PDF document that are subsequently forwarded don't work as intended. We try to explain once again what we actually want:

> **Prompt:** I want wget to locally download all files needed to correctly display the document (i.e., CSS + images) *and* to change these links within the document (one document only). But I do *not* want wget to change links to everything else. Links to external files not directly displayed within the document should stay as is. Can you help?

The prompt shows that the wget option --mirror originally suggested by ChatGPT was unsuitable. It's pointless to argue about whether we didn't describe the task precisely enough at the beginning or whether the blame lies with ChatGPT. In any case, a relatively long interactive process was required until the script produced in collaboration with ChatGPT worked as intended. The final script looks as follows:

```
#!/bin/bash
URL=$1
shift    # shift the arguments to remove the URL
TMP_DIR=$(mktemp -d /tmp/html2pdf-tmp.XXXXXXXX)
trap 'rm -rf "$TMPDIR"' EXIT
cd "$TMP_DIR"
wget --page-requisites --convert-links --no-parent "$URL"
# if $URL contains a long query string, wget shortens the file
# name; therefore, find the newest file in the result list
# with the same base name; this should be reasonably safe
URL_NO_QUERY=${URL%%\?*}
BASE_NAME=$(basename "$URL_NO_QUERY")
MAIN_FILE=$(find . -type f -name "${BASE_NAME}*" -print0 | \
            xargs -0 ls -t | head -n 1)
# run html2pdf
/usr/local/bin/html2pdf "$MAIN_FILE" "$@"
```

8.5.2   Linux Server Configuration with "tmpfs"

The script now runs very robustly on a Linux server. To improve efficiency, we've configured the server so that the /tmp directory is mapped as a tmpfs file system in the working memory. Want to know how to do this? Ask an AI tool!

> **Prompt:** On an Ubuntu server, I want to save the contents of /tmp in memory to make I/O faster. /tmp must not use more than 4 GB RAM. Please show me a systemd compatible setup.

To create a temporary file system in memory for /tmp using tmpfs and limit it to 4 GB of RAM, you can configure a systemd unit for this purpose. Here's how you can set it up:

- First, create a new systemd unit file to mount /tmp as a tmpfs with a 4 GB size limit.

  ```
  sudo nano /etc/systemd/system/tmp.mount
  ```

[more]

ChatGPT also explains the structure of the tmp.mount file (12 lines of code) and its activation via systemctl daemon-reload and systemctl start tmp.mount. Basically, this is a good solution, but it could be even better. In modern Linux distributions, /lib/systemd/system/tmp.mount already exists. It would be sufficient to make only minimal changes to the existing setup. Not until you put the AI tool on the right track does it describe this procedure:

Prompt: My systemd distribution already contains a preconfigured file tmp.mount. With the exception of the disk size, it's perfect. Can't I simply change the setup to use a maximum of *n* GB RAM?

Yes, if your system already includes a preconfigured tmp.mount unit file, you can simply modify the existing file to set the maximum size to 4 GB. Here are the steps:

- Open the tmp.mount file for editing:

  ```
  sudo systemctl edit tmp.mount
  ```

- In the override file, add the following content to set the size limit for the tmpfs mount:

  ```
  [Mount]
  Options=mode=1777,size=4G
  ```

- Reload the systemd daemon to apply the changes:

  ```
  sudo systemctl daemon-reload
  ```

- Restart the tmp.mount unit to apply the new size limit:

  ```
  sudo systemctl restart tmp.mount
  ```

8.5.3   Conclusion

This example was very close to our day-to-day administration work. Did ChatGPT help or not? For the first part of the question, ChatGPT and GitHub Copilot immediately created a basic framework for the script. This part of the task wasn't difficult, but it certainly saved us a few minutes of typing.

The real problem was finding the right mix of options for wget. We only use wget from time to time and don't know its options by heart. ChatGPT initially led us astray, but ultimately provided the right options. The alternative would have been to look for the options in the manual. This isn't easy because `wget` provides a rather confusing text, the length of which corresponds to about 40 pages of this book. So, it's hard to say whether we would have reached our destination faster this way.

For the second part of the question, the solution proposed by ChatGPT was basically good. The second variant with `systemctl edit` is preferable for modern distributions, however, because it only requires a minimal change to the setup that is already provided. However, you can only access the corresponding instructions if you already know the perfect setup and explicitly point this out to ChatGPT. But, under this assumption, you don't need an AI tool at all.

If we had to grade the whole process, it would likely be a B: the AI tools worked *well*, but not *very well*.

8.6   Calling GitHub Copilot and ChatGPT in the Terminal

If you work a lot in the terminal, it makes sense to call ChatGPT and others directly from there. There are various ways of doing this, two of which are described here: `gh` and `sgpt`.

8.6.1   GitHub Command-Line Interface ("gh")

The `gh` command enables you to carry out various GitHub actions directly in the terminal. `gh` is therefore a *command-line interface* (CLI) to the GitHub web interface. Installation instructions for Windows, Linux, and macOS can be found at *https://github.com/cli/cli*.

You must carry out authentication with `gh auth` before using it for the first time. Finally, you must install the `gh` extension, `gh-copilot`:

```
gh extension install github/gh-copilot
```

Assuming that you have a GitHub Copilot subscription, you can then call two Copilot commands:

- `gh copilot explain "command"` provides explanations of the specified command.
- `gh copilot suggest "question"` suggests a solution for the specified question. Using `-t shell|gh|git` you can explicitly refer to terminal commands (Bash, Z shell [Zsh], PowerShell), to the `gh` command or to Git problems. In the course of further steps, `gh` can explain the suggested command, copy it to the clipboard, or execute it.

The following examples show how you can use the commands:

```
gh copilot explain "sed 5,7d < file"

  Explanation:

  * sed is a command-line tool for text manipulation.
  * 5,7d specifies that we want to delete lines 5 to 7 from
    the input.
  * < file redirects the contents of the file to the input of
    the sed command.

gh copilot explain "Get-Process | Select-Object -First 5"

  * Get-Process retrieves a list of running processes.
  * Select-Object -First 5 selects the first 5 processes from
    the list.

gh copilot suggest -t shell "count IP addresses in a log file"

  Suggestion:

  grep -oE "\b([0-9]{1,3}\.){3}[0-9]{1,3}\b" logfile.txt | \
    sort | uniq -c | sort -nr

  ? Select an option
  > Explain command

  * grep -oE "\b([0-9]{1,3}\.){3}[0-9]{1,3}\b" searches for
    IP addresses in the logfile.txt file ...
```

8.6.2   Shell-GPT ("sgpt")

ChatGPT is only intended for use in web browsers. In the past, there have been tools that have tried to circumvent this. However, these no longer work today. The use of ChatGPT in the terminal therefore requires that you have paid API access to OpenAI (see *https://openai.com/api*). A ChatGPT subscription isn't enough!

Based on this condition, you have the choice between various commands that can be run in the terminal. These commands send API requests to OpenAI and then display the response. Each call costs a little less than a cent, depending on the length of the answer. In mid-2024, OpenAI charged $10 for one million output tokens generated with GPT-4o. This corresponds to a text volume of approximately 800 book pages.

We've tried sgpt. Installation instructions for Linux, macOS and Windows can be found at *https://github.com/tbckr/sgpt*.

The command requires the OPENAI_API_KEY environment variable to be defined. On Linux or macOS, you simply need to include the following statement in .bashrc or .zshrc. You must log out and log in again for the variable declaration to take effect.

```
# in the .bashrc (Linux) or .zshrc (macOS) file
...
export OPENAI_API_KEY='sk-Ke...'
```

The following listing shows three application examples. The last example is particularly interesting, where we asked sgpt to process the very confusing result of the whois command from the standard input in such a way that it's easier to understand (from a human perspective).

```
sgpt "is 172.16.5.240 in 172.16.5.160/28"

  No, 172.16.5.240 is not in the range 172.16.5.160/28.

  The network address 172.16.5.160 with a subnet mask of /28
  means that the range of IP addresses includes 172.16.5.161 to
  172.16.5.174.

sgpt "bash: how do I replace ; by : in a text file"

  You can use the `sed` command in the terminal to replace `;`
  by `:` in a text file. Here's an example command:

  sed -i 's/;/:/g' file.txt

whois kofler.info | sgpt stdin "Summarize human readable"

  The domain kofler.info is registered to Michael Kofler. The
  registrar is Key-Systems GmbH and the domain was created on
  September 26, 2003 and expires ...
```

8.6.3   Using Language Models in the Terminal ("llm")

The sgpt command just described works well, but is specifically designed for the OpenAI API. The Python program llm takes a much more general approach. It supports various APIs and can use both network and local language models. Using llm, you can change the language model at any time.

> **Note**
>
> For more information on llm, take a look at these websites:
>
> - *https://github.com/simonw/llm*
> - *https://simonwillison.net/2024/Jun/17/cli-language-models*

It's best to use pip or brew (macOS) to install the Python script. Various libraries and modules are also downloaded in this process. If you use pip, you should first set up a directory with a Python environment and activate it. In our tests, the pip installation on Linux only worked on x86 hardware, but not if using an ARM CPU. (The errors don't occur during installation, but only later when the llm command gets used.)

```
pip[3] install llm
brew install llm
```

The next step is to install the gpt4all plugin, which allows llm to communicate with various language models (including the OpenAI API):

```
llm install llm-gpt4all
llm models list

  OpenAI Chat: gpt-4o (aliases: 4o)
  OpenAI Chat: gpt-4o-mini (aliases: 4o-mini)
  ...
  gpt4all: Meta-Llama-3 - Llama 3.1 8B Instruct 128k,
           4.34GB download, needs 8GB RAM
  gpt4all: orca-mini-3b-gguf2-q4_0 - Mini Orca (Small),
           1.84GB download, needs 4GB RAM
  gpt4all: Phi-3-mini-4k-instruct - Phi-3 Mini Instruct,
           2.03GB download, needs 4GB RAM
  gpt4all: wizardlm-13b-v1 - Wizard v1.2,
           6.86GB download, needs 16GB RAM
```

If llm is to use the OpenAI API, you should install the corresponding plugin in the next step and enter your API key:

```
llm keys set openai

  Enter key: *******
```

You can then use llm to send prompts to the OpenAI API. The answer gets displayed in the terminal:

```
llm 'explain recursion'

  Recursion is a programming and mathematical concept where a
  function calls itself directly or indirectly to solve a
  problem. It's a powerful technique that can simplify code and
  make certain problems easier to solve ...
```

Provided you have powerful hardware, you can also process prompts using a local language model. Prior to that, the model should be downloaded and saved in the .cache/gpt4all directory.

```
llm -m orca-mini-3b-gguf2-q4_0 'explain recursion'

  Recursion is a programming technique ...
```

We'll leave it at that for now. You can find lots of tips on using llm on the GitHub page we gave the URL for earlier. In addition, we deal with the local execution of language models in detail in this book from Chapter 9 onward.

PART II

Local Language Models and Advanced AI Tools

Chapter 9
Executing Language Models Locally

Large language models (LLMs) have been the talk of the town since the groundbreaking success of ChatGPT. Whether it's Gemini, Claude, or ChatGPT, they all have one thing in common: they are provided by companies, and these companies even provide the interfaces to their models free of charge. As a Pro customer, you'll then receive certain features such as priority processing of inquiries or the latest and best model variants for a relatively modest fee.

Whichever option you choose, you're always dealing with a service that is hosted in the cloud, so you don't know for sure what happens to the data that you transmit to the services via your prompts. Another criticism of these online services is that you're dependent on the respective provider. Currently, a sufficient number of offers are available so that you have no problem if a provider withdraws its offer, but the dependency remains.

Concerns about privacy and dependence on one provider in particular have led some people and companies to look for alternatives when it comes to using language models. This is why there is a growing interest in executing LLMs locally. This chapter reveals how well this works:

- First, we'll provide an overview of freely available LLMs. We also address the question of how powerful your hardware needs to be in order to run language models locally at a reasonable speed.

- Then, we describe the installation and use of the free *GPT4All* and *Ollama* programs. Both programs are compatible with a wide range of LLMs and can even take local documents into account when generating the response. While GPT4All is intended for beginners, Ollama provides a good basis for many professional applications, from coding to application programming interface (API) use.

- Ollama is actually a purely text-oriented tool. However, you can combine the program with the *Open WebUI* interface and then execute prompts similar to the web-based versions of ChatGPT and others.

- The *Continue* plug-in implements a code wizard you can use instead of GitHub Copilot or other commercial services. Continue is able to communicate with a local Ollama installation or with external LLM APIs.

- Ollama can also be used via an API. This allows you to write your own programs that don't rely on commercial LLMs, but instead use a locally executed language model.

- The *Tabby* project represents an alternative to the combination of Ollama and Continue: it consists of an LLM server, which can be installed on your own computer or on a powerful workstation in the company, and an editor plug-in with code wizard functions.

9.1   Spoiled for Choice of Large Language Model

If you deal with generative AI, you'll inevitably come across various LLMs. This section is intended to give you a rough overview of the different model families, their intended use, and the resources required for local models.

9.1.1   Which Models Solve Which Problems?

In general, you need to distinguish between commercial and freely available models for LLMs. The commercial models are provided by providers as services. ChatGPT is just one example of such a service. In the basic version, you can use ChatGPT for your requests free of charge in your browser. OpenAI, the provider behind ChatGPT, offers a monthly subscription model with several benefits, such as access to better models, higher message volumes or earlier access to new features. Anthropic is pursuing similar strategies with Claude as is Google with Gemini.

In addition to the familiar browser interface or apps, you can also use these models via an API. Here the billing model is usually different, and you have to pay per token, that is, both for input and output tokens. The fees per token are usually very low. For example, you can buy one million input tokens from OpenAI for $2.50 and one million output tokens for $10. The smaller GPT-4o Mini model costs $0.15 for one million input tokens and $0.60 for one million output tokens (as of January 2025).

There are no costs for the local version of freely available models such as Llama, Gemma, or Mistral. However, you must now provide the infrastructure on which the models are executed. For this reason, you should weigh very carefully which variant you choose. You can usually control the costs through the quality of the model and the response time.

There are currently many different LLMs on the market. Some models share the same architecture and can be grouped into families. With GPT, Llama, Mistral, and Claude, you'll get to know some examples and their respective special features in the following list:

- **GPT: Generative Pre-Trained Transformers by OpenAI**
  Probably the best-known models belong to the GPT family from OpenAI. They form the basis for ChatGPT. GPT is based on the transformer model and uses a special variant of it that is optimized for text generation. A key to the success of these models is comprehensive training with a large amount of text and subsequent fine-tuning.

OpenAI provides the GPT models via its own services. Its use is partly free of charge and partly subject to charge. The models in the GPT family can be multimodal; that is, they can work not only with text but also with images or audio. Unfortunately, the LLM files from OpenAI can't be downloaded for free.

Typical tasks include chatbots, content generation, and code generation.

- **Llama: Large Language Model Meta AI**
Llama, the language models from Meta, use a similar architecture to GPT and are based on the transformer architecture. Thanks to various optimizations and design decisions, the focus of these models is more on efficiency and performance with a smaller model size.

Meta's Llama models are subject to the Llama Community License Agreement. This license is intended for private use and research, but restricts the commercial use of the models.

Typical tasks include natural language processing (NLP) tasks such as text generation, translation, and research tasks. Code Llama is a variant of Llama optimized for coding tasks.

- **Mistral/Mixtral**
The Mistral and Mixtral models are made by Mistral, a French company specializing in AI. The company is known for its open-source models, which are subject to the Apache 2 license. In addition to these, Mistral also offers other paid models via its own APIs.

Mixtral, the latest models from Mistral, are Mixture-of-Experts Language Models (MoE-LLMs) which combine several expert models and thus achieve better performance and a high degree of accuracy. In contrast to GPT and Llama, Mixtral isn't a complete transformer model, but a pure decoder model.

Typical tasks include summarizing texts, structuring, answering questions, and code completion.

- **Claude**
The Claude models are developed by Anthropic. The focus of development is not only on the performance of the models but also on safety and ethics.

At their core, the Claude models are also based on a transformer architecture. The difference lies in the constitutional AI, a special training method based on supervised learning and reinforcement learning. Another special feature of the modern Claude models is their context size of 200,000 tokens, which is very large compared to the competition. Unfortunately, like OpenAI, Anthropic doesn't offer a download option for its models.

Typical tasks include commercial applications, complex analyses, summarizing large amounts of data, and coding.

An objective comparison of which language model delivers the ideal result for which application is extremely difficult. There is no generally accepted standard for this yet, but there are numerous projects and scientific studies. Reading the following, selected with regard to coding applications, will give you a first impression:

- *https://lmarena.ai/?leaderboard*
- *https://artificialanalysis.ai/models/gpt-4o/providers*
- *https://github.com/continuedev/what-llm-to-use*
- *https://evalplus.github.io/leaderboard.html*
- *https://symflower.com/en/company/blog/2024/comparing-llm-benchmarks*

Ultimately, you'll have to experiment yourself to find out which LLM works best on your hardware and for your requirements.

Free or Truly Open Source?

"Free" LLMs such as Llama, Gemma, StarCoder, or Mistral are subject to separate licenses, which may restrict their use. Most LLMs have more restrictions than you're used to from the Linux world, for example.

Especially before you start using free LLMs commercially, you must familiarize yourself with the terms of use. These can be found on the respective project page and on *https://huggingface.co* in the description of the respective language model.

9.1.2 Hardware Requirements

Being able to download numerous language models free of charge is great, but the local execution of such models using programs such as GPT4All or Ollama often fails due to the high hardware requirements. For example, if you try to run the Llama 3.1 model in the best quality 405B version on a standard computer, it's simply not possible. For this reason, we'll now take a look at the general hardware requirements for running an LLM:

- **CPU**
  You should have at least an 11th generation Intel chip or a Zen4-based AMD CPU. Apple's M-CPUs are also very suitable. More important than the sheer number of cores is the feature set of the processors, especially the accelerated matrix multiplication required by AI models.

- **RAM**
  For a 7B model, your system should have at least 16 GB of RAM. (7B means 7 billion parameters.)

- **Hard disk space**
  You should plan about 50 GB of free space on your hard disk so that you can save a few models to try out.

■ **Graphics processing unit (GPU)**
Although a graphics card isn't absolutely necessary, it speeds up the execution of the models considerably. However, the entire model must be stored in the video memory of the graphics card. For a 4-bit quantized model, you should expect the following VRAM sizes:

– 7B: about 4 GB VRAM

– 13B: about 8 GB VRAM

– 30B: about 16 GB VRAM

– 65B: about 32 GB VRAM

Apple computers with ARM CPUs (Apple Silicon) are a special case. Their GPUs can access the entire working memory, so there is no need to differentiate between RAM and VRAM.

These recommendations are from Justin Hayes, one of the contributors to Open WebUI and Ollama, and can be found in a GitHub issue of Open WebUI here:

https://github.com/open-webui/open-webui/discussions/736#discussioncomment-8474297

As long as you provide your platform with sufficient memory so that it can load the corresponding model, you can also run it. However, if you use hardware that is too weak, you must expect the model's feedback to be very slow. It's better to use a slightly smaller model that works smoothly on your system rather than a model that can just about run on your system.

Speed Benchmarks

A decisive parameter for estimating how well the local execution of a language model works is the number of tokens that can be processed (input) or generated (output) per second. For interactive use in coding, your local language model should achieve at least 30 tokens/s (output).

The token rate depends primarily on two things: the model size and the available hardware. On the following web pages, you'll find results for various LLMs and hardware configurations on Windows, Linux, and macOS:

■ *https://llm.aidatatools.com/results-windows.php*

■ *https://llm.aidatatools.com/results-linux.php*

■ *https://llm.aidatatools.com/results-macos.php*

9.1.3 Alternatives to the Local Execution of Models

In addition to the local execution of AI models and the use of commercial models, there is another intermediate step. You can also rent virtual machines with GPUs in the cloud. All major providers offer such services. Note, however, that the use of such virtual machines can sometimes incur considerable costs, so you should carefully consider whether you only need them occasionally. The following overview shows a selection of options for such platforms from the major cloud providers:

- **Amazon Web Services (AWS): EC2 GPU Instance**
  https://aws.amazon.com/ec2/instance-types/#Accelerated_Computing
- **Google Cloud Platform (GCP): Cloud GPU**
  https://cloud.google.com/gpu
- **Microsoft Azure: NVIDIA Deep Learning (ND) or NVIDIA Compute (NC) family compute instances**
  https://learn.microsoft.com/en-us/azure/virtual-machines/sizes/overview
- **NVIDIA: GPU cloud computing**
  www.nvidia.com/en-us/data-center/gpu-cloud-computing/

The various providers offer cloud instances in different sizes. The prices depend heavily on the respective configuration and the term. For example, if you rent an *NCasT4_v3* instance of Microsoft Azure, it will cost you around $500 per month. In return, you get 28 GB RAM, 176 GB hard disk space, and an NVIDIA Tesla T4 GPU with 16 GB VRAM.

You can increase this configuration even further if you select a *NC48ads_A100_v4* instance, for example. This instance offers you 880 GB RAM, 96 Cores, 3,832 Temporary Storage, and 4 NVIDIA A100 GPUs with 80 GB VRAM each. With this configuration, however, the costs rise to around $14,000 per month!

For an exact price calculation, you should consult the price calculator of the respective provider. No matter which commercial provider you choose, there will always be costs for you. An alternative is local execution, either on your own computer or in your local network. For this purpose, you can use platforms such as GPT4All or Ollama, which make it easier for you to run models.

Groq

With its GroqCloud, the company Groq (not to be confused with the AI chatbot Grok from xAI!) is treading a middle path between commercial AI API providers such as OpenAI or Anthropic and pure cloud/hardware providers such as AWS. Groq enables you to execute LLM prompts via API keys. Various open-source LLMs in different sizes are available to choose from. Billing is based on usage (number of tokens with different prices depending on the model). Groq uses an internally developed AI accelerator (*Language Processing Unit*) that can execute LLMs particularly quickly. You can find information on this topic at *https://groq.com/groqcloud*.

9.2   GPT4All

Probably the easiest way to try out local LLMs is the open-source program *GPT4All* from Nomic AI. The aim of the project is to run LLMs on local hardware (notebooks or desktop PCs) while offering a beginner-friendly interface. Another advantage is that the hardware requirements are relatively low and the application can run on a wide range of hardware. The most important features of GPT4All include the following:

- **Privacy**
  GPT4All attaches great importance to data privacy. The application models are executed locally on your system so that all input data, processing steps, and outputs don't leave the local machine. This minimizes the risk of sensitive data being intercepted by third parties.

- **Offline capability**
  GPT4All is an open-source project designed as a desktop application for Windows, macOS, and Ubuntu. After the installation, you can install LLMs locally and use them in the graphical user interface (GUI).

- **Many models available**
  GPT4All isn't tied to a specific LLM, but supports a variety of models such as Llama, Mistral, and Nous-Hermes. You can easily select and use the models in the application interface.

- **User-friendly interface**
  The GPT4All UI is designed to be simple and self-explanatory so that you can interact with it directly without further explanation.

9.2.1   Installation and Use

GPT4All is an open-source desktop application that you can install on your system to interact with locally installed LLMs. You can download installers for Windows, macOS, and Ubuntu from *www.nomic.ai/gpt4all*. After installation, you can launch the application and use it immediately.

The core feature is the chat, which works in a similar way to the commercial products ChatGPT, Gemini, and Claude. The big difference is that you must first download and integrate a model before entering your first prompt. You can add new models either directly on the start page via **Find Models** or via the navigation under **Models**. If you use the regular navigation, you'll first see a list of the locally installed models. Click the **Add Model** button to go to the model search and install a new model.

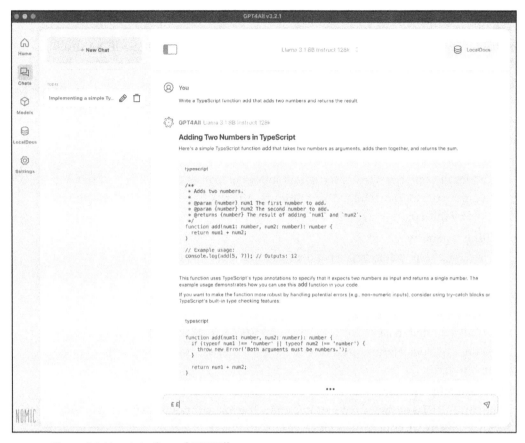

Figure 9.1 User Interface of GPT4All

GPT4All provides a brief description and the most important key figures for each model:

- **File size**
  When using local LLMs, you should have enough free space on your system's hard disk. The smaller models are in the single-digit gigabyte range, while the larger versions of Meta's Llama model already require more than 100 GB of memory. This storage space applies to the model and its weights. Larger models can depict complex relationships, but they also require more storage space.

- **RAM required**
  You need RAM to run a model. The more extensive the model, the more RAM you need to provide to avoid slowdowns or crashes.

- **Parameters**
  The parameters include the weights and biases of an LLM. They control how the model processes language and generates output. The more parameters an LLM has,

the better it's able to process complex issues and generate better output. The disadvantage of a high number of parameters is that the model becomes significantly more computationally intensive and memory-intensive.

- **Quant**
  The quantization indicates the precision of the model weights. Internally, many models work with 32-bit floating point numbers. The accuracy can be reduced by quantization. Frequently used values are 4, 8, or 16 bits. Although this reduction is at the expense of precision, it increases speed and reduces memory requirements.

- **Type**
  The model type indicates the underlying model. Possible types include LLaMA3, Mistral, or GPT.

Once you've decided on one or more models, you can download them. Once the download is complete, GPT4All will provide a dropdown menu in the chat window where you can select the desired model and then use it for the current chat session.

9.2.2   Chat with Your Files

Another interesting feature of GPT4All is hidden behind the **LocalDocs** navigation item. GPT4All allows you to integrate your local documents and use them for queries. As the application runs completely locally, your documents don't leave the system and are therefore secure.

To integrate your documents, you should create a document collection by specifying a directory on your system in which the documents are located. GPT4All processes all documents in this directory and then makes them available for queries. For a new chat, you then select the model to be used and the document collection for this chat via the **LocalDocs** button. You can then formulate your prompts as usual, and GPT4All will take into account the documents in the collection. If you get a hit, you'll also receive a link to the relevant document.

The LocalDocs feature works via embeddings (see also Chapter 12). For this purpose, GPT4All stores the data from the documents in vectors and records their meaning. With the help of these vectors, the application is able to provide answers to questions about the documents very quickly. Internally, GPT4All uses the embedding models from Nomic AI, the manufacturer of GPT4All. The application then integrates the text passages found into the prompts to make the output more accurate and informative.

GPT4All supports plain text, PDF, and Markdown as document formats. When using GPT4All in development, this means that you can, for example, integrate books on design patterns or architecture as LocalDocs, or you can use the documentation of a programming language or framework directly to generate specific answers. This also allows you to avoid the disadvantage that LLMs usually don't know the latest development status of frameworks and libraries.

9.2.3 Conclusion

GPT4All is a convenient solution for getting started with local language models. It provides you with a user-friendly interface and allows you to add new models. In addition to the models you find via the search, you can also install and use your own models. A very useful feature is LocalDocs, which allows you to easily integrate local documents into your prompts without having to worry about the underlying mechanisms.

In addition to these features, GPT4All also offers you the option of accessing the application via various language bindings, so that you can access a local language model in a Python or JavaScript application, for example.

GPT4All is intended for getting started with local language models. Sooner or later, you'll reach the limits of the application, especially in a professional environment. In the rest of this chapter, you'll learn about alternatives to GPT4All for local language models, which are somewhat more complex to set up and configure but give you better control.

9.3   Ollama

Like GPT4All, Ollama is a tool that allows you to run AI models locally on your system. In contrast to GPT4All, however, Ollama doesn't provide a GUI, but focuses on local execution and the management of models in the terminal. The application is so flexible that you can add a graphical interface or integrations into other applications on your system in just a few steps.

Ollama is an open-source project managed on GitHub (*https://github.com/ollama/ollama*). The official website of the project with admittedly somewhat sparse documentation, but an interesting blog can be found at *https://ollama.com*.

Ollama is able to make optimal use of your system's hardware and can use both the system's CPU and GPU to run the models. You should give preference to using the GPU, as the performance is significantly better. The platform supports a range of NVIDIA and AMD graphics cards as well as GPU acceleration on Apple devices. An overview of the supported graphics cards can be found at *https://github.com/ollama/ollama/blob/main/docs/gpu.md*.

9.3.1   Benefits of Ollama

Ollama is a lightweight and flexible platform that allows you to run your local models. There are several reasons why you should use this platform:

- **Local execution**
  With Ollama, as with GPT4All, you're independent of AI solution providers, as all elements of the platform, including the models, run locally on your system. In addition

to this independence, local execution also reduces communication latency so that you receive your answers faster.

- **Simple operation**
  Using the Ollama command-line tool, you can carry out all the necessary operations directly with a straightforward set of commands. Although this variant isn't particularly convenient, it allows you to run Ollama on a system without a GUI and control it via a remote connection.

- **Preconfigured models**
  Ollama provides you with a range of predefined models that you can load directly via the command line and integrate into your system.

- **Customizability**
  Loading the predefined models is the most convenient way to get started quickly with frequently used, freely available models. However, Ollama also allows you to integrate your own models.

- **Flexibility**
  In the Ollama tool, freedom and flexibility run through everything from the models to the UIs. You can design the platform entirely according to your needs with hardly any restrictions.

- **Privacy**
  Local execution means that your prompts and responses don't leave your system. If you run Ollama locally, your data is protected against unauthorized access. If you operate the platform in your local network or even publicly available on the internet, you must protect the interfaces against misuse.

- **Cross-platform**
  You can run Ollama on different systems. There are installation routines for macOS, Linux, and Windows. You can also operate the platform in a Docker container.

Now that you know the main advantages of Ollama, let's take a closer look at the general structure of the platform and its individual components. Understanding how Ollama is structured will help you better understand how it works and the customization options that are available.

9.3.2   General Structure of Ollama

Ollama tries to make it as easy as possible to start working with local models, so that you have a working environment in just a few steps. Ollama consists of several parts that work together to create a functioning local environment:

- **Ollama engine**
  The core of Ollama is the engine that is responsible for running the AI models. It allocates resources and optimizes execution. The engine runs as a service in the background and provides the interfaces to the model.

- **Command-line tool**

  The easiest way to interact with Ollama is via the command line. This tool allows you to manage the local models and also send prompts.

- **Local model**

  The Ollama platform alone is of little use without at least one local model installed. You can use the command line to load one of the models from the Ollama library or integrate your own model.

You must install these components on your system in the next step to be able to use Ollama.

9.3.3 Local Installation and Execution

There are various options available for installing Ollama. The first port of call is the official website (*https://ollama.com*) where you get the Ollama application for macOS, which you can run immediately. An installation script is provided for Linux, which you can download via an HTTP client such as curl and pipe into your shell. A setup file is available for download to Windows.

Alternatively, you can also use your system's package manager. On macOS, you can install Ollama with Homebrew, and on an Ubuntu system, you can install it using snap.

After the installation, a range of command-line options are available to you. You can test the general functionality on the command line via the following command:

```
ollama -v
  ollama version is 0.3.6
```

Before you can work with Ollama, you must install a local model. The following models are available in the Ollama library:

- **Llama 3.1**

  A powerful language model from Meta designed specifically for NLP and conversational AI. It provides improvements in text generation and text comprehension compared to its predecessors.

- **Phi 3**

  An open LLM from Microsoft that delivers very good performance with a relatively small size and performs excellently in various benchmarks in language processing, logic, programming, and mathematics.

- **Gemma 2**

  A generative model from Google that is characterized by high efficiency, performance, and accessibility. It was developed primarily for developers and researchers.

- **Llama 2 Uncensored**

  A variant of Llama 2 that is less restrictive in terms of content moderation and can therefore be used more flexibly in content generation, especially in contexts where less controlled text output is required.

- **Mistral**
  A model that specializes in the recognition and processing of multilingualism. It's designed for applications that have to work in several languages and require a high degree of accuracy in language processing.

- **Starling**
  A model that specializes in precision in the reproduction of information. It's often used in applications that require reliable and accurate information, such as question-and-answer systems or knowledge databases.

- **Code Llama**
  A specialized model from Meta for code generation and completion. It supports multiple programming languages and is particularly useful for developers looking for support with code creation and optimization.

- **LLaVA**
  A model that combines multimodal capabilities, such as the processing of text and visual data. It's ideal for applications that require the integration of image and text information such as image descriptions or visual question-and-answer systems.

For some models, such as Llama 3.1, there are different variants available. These range from the smallest version with 8 billion parameters and 4.7 GB to the medium version with 70 billion parameters and 40 GB to the largest version with 405 billion parameters and 231 GB. If there are several variants, which is reflected in the name. For example, the large Llama 3.1 model is called `llama3.1:405b`. For the first steps, we install the 8B version of Llama 3.1 by using the following command:

```
ollama pull llama3.1
```

```
pulling manifest
pulling 8eeb52dfb3bb... 100% 4.7 GB
pulling 11ce4ee3e170... 100% 1.7 KB
pulling 0ba8f0e314b4... 100%  12 KB
pulling 56bb8bd477a5... 100%   96 B
pulling 1a4c3c319823... 100%  485 B
verifying sha256 digest
writing manifest
removing any unused layers
success
```

By using `ollama list`, you can display the locally available models. You'll receive information such as the name of the model, its unique ID, the size, and the date of the last change. Depending on the system, Ollama saves the models either in a subdirectory of the current user (macOS and Windows) or in a subdirectory of /usr/share (Linux).

The `ollama serve` command starts the Ollama service, and you can run a specific model via `ollama run`. By using `ollama run llama3.1`, you start an interactive session on the command line in which you can send any prompts, such as the following:

Prompt: Create a TypeScript function that adds two numbers.

Here is a simple TypeScript function that takes two numbers as parameters and returns their sum:

```
/**
 * Adds two numbers together.
 *
 * @param num1 The first number to add.
 * @param num2 The second number to add.
 * @returns The sum of the two numbers.
 */
function addNumbers(num1: number, num2: number): number {
  return num1 + num2;
}

// Example usage:
console.log(addNumbers(5, 7)); // Outputs: 12
```

[explanation of the code]

The result of the prompts is impressive, even with the small Llama 3.1 model. The larger the model, the better the feedback. You also have the option of getting even more out of your LLM through fine-tuning or retrieval-augmented generation (RAG). You can find out more about this topic in Chapter 12.

As the models are sometimes quite memory-intensive, there will come a time when you want to remove a locally installed model. In this case, you should rely on the command-line tool and not attempt to delete the model files from the hard disk yourself. For example, to remove the Llama 3.1 model, you can use the `ollama rm llama3.1` command. This command ensures that the model and all associated metadata gets properly deleted.

9.3.4   Running Ollama in a Docker Container

The local installation of Ollama has a decisive disadvantage: it interferes with your system. While Ollama is a pretty solid tool to work with, if you install every new AI tool on your system, it can quickly become confusing. An alternative to this is to run Ollama in a container. This means you remain independent of your local system and can even use the container on other systems, be it on other client systems or servers.

To use the Ollama container, you need the Docker engine or an alternative such as Rancher Desktop on your system. If this requirement is met, you can use the official Ollama image:

```
docker run -d -v ollama:/root/.ollama -p 11434:11434 \
  --name ollama ollama/ollama
```

This command starts a container in the background (-d), which is based on the official Ollama image (ollama/ollama). The ollama volume is mounted to the /root/.ollama path in the container. By default, the Ollama API is linked to port 11434. The -p option allows you to release this port to the outside world so that you can access it from your system. You also assign the ollama name to the container to make it easier to address.

Depending on whether you have a GPU from NVIDIA or AMD in your system, you'll need to take additional steps to enable support for your GPU. You can find detailed instructions at *https://hub.docker.com/r/ollama/ollama*.

When you run the command, Docker first searches for a local version of the image. If that isn't available, Docker downloads the image from the central registry and creates and starts the container. As soon as the container has been started, you can interact with it.

Usually, you use the docker exec -it ollama command followed by the actual Ollama command. This Docker command executes the specified command in the running container named ollama. The i option stands for an interactive session, and t stands for TTY, a pseudo-terminal. Using docker exec -it ollama I run llama3.1, you start an interactive command-line session with the Llama 3.1 model. Via the locally mounted volume, Ollama downloads the models to the Ollama directory of your host system so that the data is independent of the container.

9.4   Open WebUI for Ollama

Regardless of whether you want to use Ollama for general prompts or for use in development, the command line is probably not the most convenient way. For this reason, there are a number of options for equipping the platform with an appealing and user-friendly frontend. One of the most popular ones is *Open WebUI*, formerly known as Ollama WebUI.

As the name change suggests, Open WebUI was originally developed as a frontend for Ollama. However, the project supports not only Ollama but also any LLM runner with an OpenAI-compatible API. Open WebUI provides a number of functions that make it interesting both for local operation and for installation in companies:

- **Easy installation**
  The core idea is that you host Open WebUI yourself. Like Ollama, Open WebUI also provides numerous installation options. Probably the simplest and cleanest method

is to use the official Docker image of the project. But we'll return to the installation later.

- **Modern UI**
  The UI is very similar to the ChatGPT web interface and is delivered as a responsive progressive web app (PWA). This means that you can use the website on any device, and it adapts to different resolutions. You can also install the website on your device so that you no longer notice any difference between a native app and the website.

- **Creation of models**
  Open WebUI provides you with a graphical interface that enables you to generate user-defined models for Ollama.

- **Local RAG integration**
  Similar to GPT4All, Open WebUI allows you to integrate your own documents into your prompts via RAG.

- **Web search for RAG**
  In addition to the RAG feature with local documents, you can also give your local AI access to the internet and enrich the results of the queries with the search results from the internet.

- **Image generation**
  Open WebUI supports the integration of image generators such as ComfyUI or DALL-E.

- **Cross-model chats**
  You can combine several models in one prompt at the same time and use their individual strengths.

- **Role-based access control**
  One feature that makes Open WebUI particularly interesting in a larger context is the support of role-based access control (RBAC). This feature allows you to grant certain people access to the UI and exclude others.

9.4.1 Installing Open WebUI

In the documentation, Open WebUI recommends installing it as a container. The prerequisite is that you've installed a container platform on your system. There's a Docker image for Open WebUI that you can start as a container via the following command:

```
docker run -d -p 3000:8080 \
  --add-host=host.docker.internal:host-gateway \
  -v open-webui:/app/backend/data \
  --name open-webui \
  --restart always \
  ghcr.io/open-webui/open-webui:main
```

Once the container has been started, you can access the Open WebUI web interface via *http://localhost:3000*. The user login feature is activated by default. If you want to use the tool in single-user mode and do without a login, you can deactivate this feature using the WEBUI_AUTH environment variable with the False value. If you use the standard mode, you'll first see a login screen when you open the UI for the first time. You can create an initial user account via the **Sign up** link, which is automatically activated and has admin privileges.

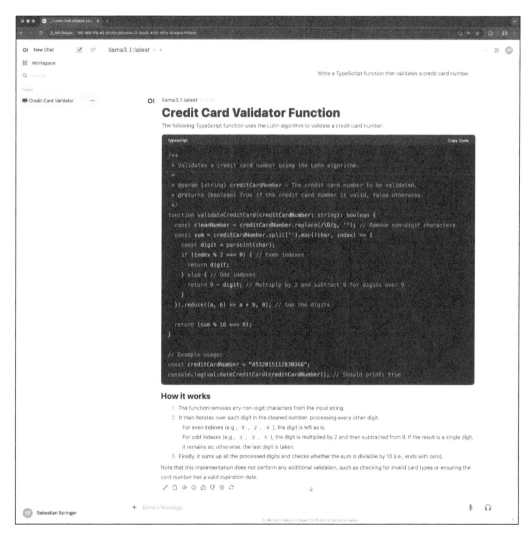

Figure 9.2 Web Interface of Open WebUI

Before you start prompting, you must make sure that Ollama is started and accessible as a backend. If you run both applications in separate Docker containers, communication usually works without any further adjustments. You can then select one of the models provided by Ollama via the **Select a model** dropdown menu. By clicking on **Set as default**, you set this model as the default for all future sessions and don't have to select it again each time.

With the combination of Ollama and Open WebUI, you're independent of AI providers such as OpenAI, Google, or Anthropic. Your data remains on your system, and you can control the performance of your system depending on how many local resources you invest. Models such as the 405B variant of Llama 3.1 have no reason to hide behind the large commercial providers, but they require an extremely large amount of storage space, RAM, and computing power.

Thanks to the GUI of Open WebUI, operating the services is very convenient. In the following sections, you'll learn more about the possibilities provided by the combination of Ollama and Open WebUI.

9.4.2 Using Custom Documents in Open WebUI

If you use LLMs for development, the results of prompts on general questions are particularly impressive. However, the more specific the questions get, the greater the probability that the AI tool will be wrong. In the best-case scenario, it admits this and tells you that it has reached the end of its knowledge. However, it's more likely that you'll simply receive the wrong answer.

By using RAG, however, you can help the AI tool and convey specific specialist knowledge to it. You've already become acquainted with a variant of this with the LocalDocs of GPT4All, and Chapter 12 of this book deals with this topic in even greater detail. Open WebUI also provides the option of uploading your own documents and enriching the answers to your prompts with them. For example, you can upload manuals and documentation for frameworks, libraries, or programming languages; prompt specific issues; and receive a suitable answer.

To use RAG in Open WebUI, you must first upload one or more documents. To do this, navigate to **Workspace**, and select the **Documents** tab. Initially, you'll see an empty list here, as no documents are available yet. Documents can be uploaded by using the plus button. Open WebUI uses tags for documents and collections. Each document is given an individual name tag that you can use to reference it. You can also assign any number of tags to each document. Documents with identical tags are combined into collections. As soon as you've uploaded at least one document, you can enter the hash (#) as the first character in a prompt. Open WebUI then gives you a selection of the available tags whose documents you can integrate into the current session via a dropdown menu.

The RAG feature of Open WebUI supports a variety of file formats, including text files such as source code files, Markdown, XML, Microsoft Excel, PDF, or EPUB. Internally, Open WebUI processes these files via various loaders.

You can configure the RAG feature under **Documents** in the admin panel, which you can access via the admin user settings. There you specify which embedding model engine you want to use. The default engine is **SentenceTransformers**. This is a Python library that converts sentences and text fragments into vectors (embeddings). The RAG is then based on these vectors.

The RAG functionality of Open WebUI isn't limited to local documents. Instead of a tag that references a document, you can also enter a URL to a website in the prompt after the hash (#) and select the suggested link. The application then reads the page and uses its structure to generate a response. An example of such a web-based prompt could read as follows:

> **Prompt:** #https://developer.mozilla.org/en-US/docs/Web/JavaScript/Reference/Global_Objects/TypedArray
> Is it possible to freeze a `TypedArray`?
>
> No, it is not possible to freeze a `TypedArray`. According to the rules, `TypedArrays` that aren't empty cannot be frozen.

The answer will provide you with additional information about where the information was found in the source.

If you reference web pages, you should make sure that they contain as much structured text as possible because Open WebUI processes and uses this text for the answers. Ideally, the website should provide a separate reading view, as this is best suited to the web-based RAG.

The quality of the results depends heavily on the model used. If you use a small model with a small number of parameters, such as the 8B variant of Llama 3.1, you'll obtain significantly poorer quality results than with the 405B variant of the model.

The ability to search in documents and on websites is particularly helpful when you look up documentation, and it can speed up your development process considerably. The risk of receiving incorrect answers is also lower, as Open WebUI also provides references, and you can check the relevant text passage in the documentation.

9.4.3   Custom Workspace Models

The LLMs play a prominent role in both Ollama and Open WebUI; without them, both tools are virtually useless. It's therefore of great importance that the models are managed properly. Open WebUI makes it easier to use the models by providing a GUI through which you can perform various operations. You can access model management via the navigation in the left-hand column of the interface and the **Workspace** item. The first tab of the workspace is dedicated to models. There, you can see the list of models installed in the connected Ollama by default. You can clone, edit, or delete an existing model. You can also create a new model or download one of the community models.

If you want to integrate an entirely new model, this doesn't work via the workspace. Instead, you need to go to the admin panel, select the **Settings** tab and then **Models**. You'll then be able to download one of the official models from the Ollama library via **Pull a model from Ollama.com**. For example, you can enter phi3:medium, and Open WebUI will cause Ollama to download the Phi3 model in its 14B variant. Once this process is complete, you can use the model in both Ollama and Open WebUI.

Another way to integrate a new model is via the *Grokking Generative Universal Format* (GGUF) file of a model, which you can download. GGUF is a file format for storing and exchanging LLMs. Currently, the import of GGUF files is still hidden behind the **Experimental** item in the **Model** section of the admin panel. If you have the GGUF file of a model, for example, from Gemma2, locally on your system, you can upload this file using the **Upload** button and integrate the model into Ollama. This variant is particularly interesting for models that aren't part of the Ollama library. A list of ready-made GGUF files you can use can be found on Hugging Face at *https://huggingface.co/models?search=gguf*.

No matter how you integrate a model into Ollama, Open WebUI allows you to customize it for your purposes. To do this, you either edit an existing model in the workspace or create a new model there based on an existing Ollama model. You can use the following fields:

- **Name** and **model ID**
  These two fields identify your model.

- **Base Model (From)**
  This is where you can select the underlying model. All models installed on your local Ollama are available for selection in this dropdown menu.

- **Description**
  Here, you can enter a description that describes your user-defined model in more detail.

- **System prompt**
  You can use the system prompt to store statements that aren't visible when interacting via the chat interface, but which influence the behavior of the model. Examples of such system prompts include the following:
  - Use formal language.
  - Ignore certain requests.
  - Prioritize technical accuracy.

- **Advanced Params**
  This section gives you very fine control over various aspects of the model. Here, for example, you can configure the sampling temperature and thus specify whether the output should be more random or more focused, or set the size of the context.

- **Prompt Suggestions**
  You can use the prompt suggestions to store one or more suggestions for prompts. An example of such a prompt suggestion is "How do I implement a debounce function in JavaScript?" These suggestions get displayed when you start a new session with your model.

- **Knowledge, Tools, Filters, Actions**
  These items refer to Open WebUI-specific extensions for your model:
  - **Knowledge** stands for local documents that you want to make available to your model by default. This allows you to create a model that specializes in clean code and architecture. In this case, you can integrate suitable documents.
  - **Tools** stand for various aids for your chats, such as web search or API interactions.
  - **Filters** can be used to apply actions to the messages of your users or the model. For example, you can prevent certain messages, automatically translate messages, or limit the number of messages.
  - **Actions** stand for buttons in the message bar for your end users. These buttons allow your users to perform certain actions such as formatting code.

- **Capabilities**
  You can select the **Vision** option here, which you can activate or deactivate. The option stands for the model's ability to process and interpret visual information.

- **Tags**
  You can use the tags to add compact meta information for your model.

You can export the configuration of your model as a JavaScript Object Notation (JSON) structure to create a backup or share the preconfigured model with other people. The following code is an example of such a model export:

```
[
  {
    "id": "code-assistant",
    "name": "Code Assistant",
    "object": "model",
    "created": 1723984665,
    "owned_by": "ollama",
    "info": {
      "id": "code-assistant",
      "user_id": "befb1b0f-171a-41d8-b955-5283e2e45582",
      "base_model_id": "llama3.1:latest",
```

```
      "name": "Code Assistant",
      "params": {
        "system": "You are a Code Assistant, a developer's expert
        tool. Provide clear, best-practice code, technical
        advice, and debugging help for languages like JavaScript,
        TypeScript, and Python. Keep responses concise and
        implementation-ready.",
        "num_ctx": 4096
      },
      "meta": {
        "profile_image_url": "/static/favicon.png",
        "description": "A model that is specialised to assist
        you in your daily coding work.",
        "capabilities": {
          "vision": false
        },
        "suggestion_prompts": [
          {
            "content": "Generate a REST API endpoint in Node.js
            using Express."
          },
          {
            "content": "How do I optimize a SQL query for faster
            performance?"
          },
          {
            "content": "Write a Python script to parse JSON
            data."
          }
        ]
      },
      "updated_at": 1723984665,
      "created_at": 1723984665
    },
    "preset": true,
    "actions": []
  }
]
```

The big advantages of the combination of Open WebUI and Ollama over the popular online services is that your data doesn't leave your system, and you have complete control over the entire environment and all aspects of the model and interaction. As you've seen from the model configuration example, you can intervene very deeply in the system and configure it entirely according to your requirements.

9.5 Continue

For direct support in software development, the integration of AI functions directly into the development environment is the most convenient solution. There are numerous solutions for VS Code and the JetBrains integrated development environments (IDEs), of which GitHub Copilot is the best known.

There are also various alternatives with similar features, such as Bito, Cody, or Codeium. Most of these services offer a limited free version. However, if you use it on a daily basis and want to integrate the tool into your development process, you'll need the paid version in most cases. In addition, there is the problem that you need to send your data to a third-party provider.

But there's also another way! Both the *Continue* plug-in, which we present in this section, and the *Tabby* project (Section 9.7) implement a code wizard based on a locally executed LLM. This allows you to retain control over your data.

9.5.1 Installation

The Continue plug-in is available for VS Code and the JetBrains IDEs. After the installation, Continue connects to a language model of your choice, whereby local LLMs are also supported with Ollama. A quick start guide and further information on Continue can be found on the project's official website and in the Ollama blog:

- *www.continue.dev*
- *https://ollama.com/blog/continue-code-assistant*

The first step is to install the extension. Continue is currently available for VS Code and/or one of the JetBrains IDEs. When you open the extension for the first time, you can choose between three different variants:

- **Free trial**
  Here you log in with your GitHub account and receive 50 free requests. These are routed via a Continue proxy server and executed with their API keys. This variant is well suited for initial tests.
- **Local models**
  This mode requires that you host your AI models locally and, for example, operate your own Ollama installation with associated models.
- **Best experience**
  The third variant is a suggestion from the Continue team. The best commercial models are suggested here, but you'll need an API key for each of them.

To demonstrate the features of Continue, we use the VS Code extension in combination with a local installation of Ollama. In the default installation, Continue assumes the following models:

- **chat**
  Initially, Continue uses Llama3 as the model for the chat feature. The documentation recommends a model variant with more than 30B parameters if possible, but this means that the system on which you're running Ollama must have the appropriate resources. As an alternative to Llama3, Continue recommends the open-source model DeepSeek Coder v2 or the commercial models Claude 3, GPT-4, and Gemini Pro.

- **autocomplete**
  The StarCoder 2:3b model is the default setting for auto-completion. Continue recommends a model with a parameter number between 1B and 15B. Alternatives are the commercial model codestral and DeepSeek Coder v2 as an open-source version.

- **embeddings**
  Continue uses nomic-embed-text for embeddings. Alternatively, you can use the commercial model, voyage-code-2.

All settings are saved in the extension's `config.json` file and can be changed at any time. Here, for example, you can configure the various models by defining the display name and specifying where the model comes from, for example, from Ollama, as well as the name of the model.

9.5.2 Usage

After activating Continue, the extension suggests extensions or code fragments when you enter code, which you can accept or ignore. You can use the key combination `Ctrl`+`I` or `cmd`+`I` to modify the code in the currently open file. Let's suppose you have a code section where a character string is returned that is currently in German but should be translated into English. Simply place the cursor in the corresponding line, press the key combination, and formulate the prompt, *Change this value to English*. Continue then suggests a translated character string that you only need to accept.

The Continue extension chat is a versatile tool. In the simplest case, you can use a prompt to ask any question, such as `What are the parameters for Array.splice`. The model behind Continue responds with a detailed explanation, which is displayed directly in the chat window. The chat also has two special functions:

- **Slash commands**
  If you start the message with a slash (/), you'll receive a selection of predefined commands. You can use these commands, for example, to edit the selected code, add comments, or generate a commit message. Continue also provides the option of creating custom commands. If you select the last item in the dropdown (**Build a custom prompt**), you'll receive a template for such a prompt. This feature is particularly useful for tasks you have to perform frequently.

- **Context**

  You can use the @ sign to add additional context to the chat. This information helps to improve Continue's response by placing the prompt in the appropriate context.

  The context feature of Continue makes the extension a useful tool in everyday work. The ability to ask questions about your own code in the chat and at the same time integrate external sources of information means that you can receive very specific answers. The built-in context providers include the following:

  - **Code:** This context refers to specific functions or classes within your project.
  - **Git Diff:** You can use the @diff context to refer to the changes in your current Git branch and have them explained.
  - **Terminal:** The terminal provider refers to the content of the integrated command line of your development environment.
  - **URL:** You can use this context to link to external websites. Continue loads the page, converts it to Markdown, and passes it to the model as context.
  - **Jira Issues:** You can also refer to Jira issues and thus directly include the requirements of your project in your prompts or responses.

The quality of Continue's responses depends heavily on the models used and the available infrastructure. With its large commercial models, Continue is an ideal tool for everyday work. However, Continue also returns very good results in combination with Ollama, provided that sufficient storage space and computing power are available; otherwise, the responses are very slow, which makes the tool uninteresting for the development process. The decisive advantage of local execution is that your data doesn't leave your system, and you can integrate the Continue extension very deeply into your project using the context providers.

9.6   Ollama API

Tools are always connected to the Ollama platform via standardized REST interfaces, regardless of whether Open WebUI or Continue is used. These can be used not only via tools but also directly in your own applications.

Ollama provides a choice of two REST APIs that you can use to connect both your own applications and AI-supported tools. You can either use Ollama's own API (see the following section) or another REST interface that is compatible with the OpenAI API. The latter is currently still in an experimental phase and should enable you to use tools that use the OpenAI API locally in the future.

9.6.1   Ollama REST API

If you run Ollama locally on your system, either directly or in a container, you can access the platform via the address, *http://localhost:11434*. You can communicate with the Ollama API via an HTTP POST request to the /api/generate path. You can access the OpenAI API via the /v1 path. In the following code, you can see how to use the Ollama API with Node.js. As this is a simple HTTP communication, you can also implement the example on any other platform.

```
const response = await fetch(
  "http://localhost:11434/api/generate",
  {
    method: "POST",
    body: JSON.stringify({
      model: "llama3.1",
      prompt:
        "Create a TypeScript function that adds two numbers",
    }),
  }
);

const reader = response.body.getReader();
const decoder = new TextDecoder();

while (true) {
  const { value, done } = await reader.read();

  if (done === true) {
    break;
  }

  const payload = JSON.parse(
    decoder.decode(value, { stream: true })
  );

  process.stdout.write(payload.response);
}
```

The basis for this short example is an HTTP POST request to *http://localhost:11434/api/generate*. As a body, you pass an object which allows you to control Ollama and the model used. The two most important properties are model, which you use to define the model used (in this case, Llama 3.1), and prompt, which contains the request to the model. In addition, you can, for example, deactivate the streaming used by default via stream: false.

Streaming, that is, the transmission of individual tokens per package, makes communication with the model somewhat more difficult. However, it also ensures that you receive the answer more quickly, namely as soon as the model has generated the token. To consume the individual tokens and output them to the console, you want to create a reader object with the `getReader` method. You can read a package in an endless loop as soon as it's available with the `reader.read()` method. In this context, the `fetch` API uses binary data, which you have to convert into plain text using the `TextDecoder`. You can write the token generated by the model to the console of your system using `process.stdout.write`.

The save the code in a file named `api.js`. You can then execute the script via `node api.js` and receive the model's response word for word.

In addition to the individual query via the `/api/generate` endpoint, the Ollama API also provides other endpoints:

- `/api/chat`: You can start a chat session with the model via this endpoint.
- `/api/create`: Ollama allows you to create a new model using a model file.
- `/api/blobs/<digest>`: You can use this endpoint to check whether the blob file for creating a new model exists on the server. You can also create a new blob file via this endpoint (`POST`).
- `/api/tags`: Ollama returns a list of locally available models via this endpoint.
- `/api/show`: You can use this path to display additional details about a model with a `POST` request.
- `/api/copy`: This endpoint allows you to copy an existing model.
- `/api/delete`: You can use this endpoint to delete an existing model.
- `/api/pull`: You can use this endpoint to load a model from the Ollama library onto your local server.
- `/api/push`: If you're registered with Ollama.ai, you can also upload a model via the Ollama API.
- `/api/embed`: This endpoint enables you to create embeddings from a model.
- `/api/ps`: This endpoint returns a list of the models that are currently loaded into the memory.

As you can see, you can perform all operations via the REST API that you can also perform via the Ollama command line. If you operate Ollama as a server on your local network, you should note that no authentication is required to use the API; that is, the interface is open to everyone. To avoid potential problems, you should install an authentication server that only allows authorized access.

9.6.2   Libraries for Accessing the Ollama API

As direct access to the Ollama APIs can be somewhat cumbersome, there are two official libraries to help you with this: ollama-js for JavaScript and ollama-python for Python (see Chapter 10, Section 10.2). These libraries allow you to perform all operations that the REST API also offers, but in the more convenient form of functions.

The following listing shows sample code for ollama-js to submit a prompt:

```
import ollama from "ollama";

const response = await ollama.generate({
  model: "llama3.1",
  prompt: "Create a TypeScript function that adds two numbers",
  stream: true,
});

for await (const chunk of response) {
  process.stdout.write(chunk.response);
}
```

ollama-js works both on the server side with Node.js and in the browser. To use the library, you must first install the package via the npm install ollama command.

By default, ollama-js doesn't use streaming, so you can access the return directly via response.response. Alternatively, as in this example, you can activate the streaming feature and consume the content via an asynchronous for loop.

The advantage of using the libraries provided by Ollama is that your code becomes much more compact, and you have fewer details to worry about.

9.7   Tabby

Tabby (*https://tabby.tabbyml.com*) is a self-hosted AI code wizard that consists of a server and an editor plug-in. The project is an exciting alternative to the combination of Ollama and Continue. Tabby is particularly suitable if a central server is supposed to provide code wizard services for multiple employees of a company or organization.

You can run the associated server either locally on your system or centrally in your company network. Tabby itself doesn't provide an LLM, but rather allows you to choose from a whole range of established LLMs specializing in development, such as StarCoder, Code Llama, or CodeGemma. The models are dimensioned in such a way that they can be run on a reasonably powerful system.

9.7.1   Installing the Tabby Server and the Plug-In

You typically operate Tabby in a container. The project provides a Docker image for this purpose so that you can get started straight away. If you have an NVIDIA graphics card, you can take advantage of the Compute Unified Device Architecture (CUDA). All you need to do is install the NVIDIA container toolkit. Then, start the container using the following command:

```
docker run -it --gpus all \
  -p 8080:8080 -v $HOME/.tabby:/data \
  tabbyml/tabby serve --model StarCoder-1B \
  --chat-model Qwen2-1.5B-Instruct --device cuda
```

Use the following command for CPU-based execution:

```
docker run --entrypoint /opt/tabby/bin/tabby-cpu -it \
  -p 8080:8080 -v $HOME/.tabby:/data \
  tabbyml/tabby serve --model StarCoder-1B \
  --chat-model Qwen2-1.5B-Instruct
```

The `--model` and `--chat-model` options allow you to specify which model you want to use for code completion and chat, respectively.

As an alternative to running Tabby in a container, you can also install it natively on Windows, Linux, and macOS. The necessary steps can be found in the Tabby documentation at *https://tabby.tabbyml.com/docs/welcome*.

As soon as you run the Tabby server, you must create a new account, which you'll need later to connect to the server. You can access the Tabby web interface by default via *http://localhost:8080*. The first user you register here is automatically the admin user for the server.

You must invite all other users who are also supposed to use this server via the admin user settings. For this purpose, you need to enter the user's email address. As Tabby isn't connected to a mail server in its standard configuration, no email with the invitation code will be sent. However, you can copy the link to the invitation containing the code in the admin interface and send it manually to the person. This person must then complete their profile details and set a password.

The last step before you can use Tabby is to install the Tabby plug-in for your development environment. The plug-in is currently available for VS Code, IntelliJ, and Vi Improved (VIM)/NeoVIM.

9.7.2   Usage

As a concrete example, let's take a look at the Tabby integration in VS Code. After the installation, the plug-in prompts you to enter your token. This is available on the home page of the Tabby server. As soon as you've inserted and confirmed the token, the extension is ready for operation. You can use various supports in the extension:

- **Code completion**
  While you're programming, Tabby suggests suitable completions. If the suggested code fits, you can accept it by pressing the ⌜Tab⌝ key. In your daily work, Tabby often provides useful results here, which makes code completion one of Tabby's most important features.

- **Commands**
  Via the command palette of the IDE, Tabby provides you with a series of commands such as `Tabby: Generate Docs` or `Tabby: Explain this`. To use such a command, you select a code and activate the corresponding command. The result is usually displayed in the chat view of the extension.

- **Code edits**
  Code edits are a special type of Tabby command. You can use this command to have Tabby modify your source code. For example, you can select a function in a JavaScript file and give Tabby the instruction `Export this function`. The extension then inserts the appropriate `export` statement. In contrast to code completion, using code edits requires a little more patience, as the tool occasionally provides useless or empty answers. However, this feature provides great potential to catch up with the competition.

- **Chat**
  In the chat, you can interact directly with Tabby's chat LLM and ask questions that will be answered directly.

Another exciting feature is the context in Tabby. You can specify a Git repository here, for example. Tabby processes the code from this repository and saves it in its internal index. The tool then uses this information for the extension's features such as code completion or contextual queries.

9.8   Conclusion

The use of LLMs increases your flexibility to a large extent: You can choose between commercial models provided by providers such as OpenAI or Anthropic. Depending on the contract model, these providers guarantee that your data is protected against unauthorized access.

If you don't trust this guarantee or want to experiment with other language models, you can run free language models locally on your PC, on a central server in your company, or in a cloud instance with GPU support. The latter two variants are advantageous if the computing power of the company notebooks is insufficient for the LLM version.

Regardless of where the language model is executed, you can use it as a chatbot, via plug-in as a code assistant, or via API in your own applications.

In the open-source sector, you can choose from a wide range of models. Depending on the architecture and size, however, their implementation is very resource-intensive. The smaller versions, such as Llama 3.1-8B, can be run on more powerful desktop systems with a separate graphics card. For the larger models such as Llama 3.1-70B or even the 405B variant, however, you need very powerful hardware. Running such a large model is usually not worthwhile for individuals or small companies, as the infrastructure is disproportionately expensive. In this case, you return to one of the commercial providers.

In our tests, we've found that LLMs which run locally on private notebooks or PCs can't keep up with commercial cloud offerings. This is particularly true for the coding application: the token speed is too low, and the quality of the answers is noticeably lower. We expect this to change in the future: first, future computers will provide better hardware requirements for the execution of LLMs; and second, intensive research is currently being carried out to reduce the size of LLMs while maintaining the same quality.

9

Chapter 10
Automated Code Processing

So far, we've assumed that you use AI tools interactively: you ask ChatGPT for coding support or let GitHub Copilot, Continue, or another wizard help you in your integrated design environment (IDE). This is the right approach for developing new code.

But let's say you have a code base that consists of dozens, hundreds, or even thousands of files. Wouldn't it then be nice if you could formulate prompts like in the following examples?

- Change the code base from Python 2 to Python 3.
- Translate the comments in all files from German to English.
- Replace nonproportional fonts in all CSS and HTML files uniformly with "Liberation Mono".
- The schema of the `mydb` database has been changed. In table `x`, column `y` has been given the new name `z`. At the same time, the data type was changed from `INT` to `FLOAT`. Adjust all relevant code files.

Current AI tools and the associated user interfaces, editors, and development environments are (still) unable to cope with such prompts. However, this doesn't mean that such far-reaching wishes are completely unfulfillable.

This chapter begins with a crash course in the application programming interfaces (APIs) of OpenAI and Ollama. This enables you to send prompts to a language model in a separate script and analyze its responses. On this basis, we'll then show you some sample scripts that you can use to apply prompts to an entire directory with files.

Dream versus Reality

In this chapter, we've reached the limits of what is currently technically possible. Yes, via the detour of AI APIs, it's indeed possible to make automated code changes that extend across a large number of files. And no, it didn't really work very well in any of our tests. We'll document the restrictions in detail. However, as AI tools and their APIs are being developed at a breathtaking pace, what is impossible today will be within reach tomorrow and probably a reality the day after tomorrow!

That said, it hopefully goes without saying that you can't adopt far-reaching code changes now or in the future without testing them. It's advisable to make the changes in at least two Git commits. The first commit contains the changes made by the AI tool, and the subsequent commits contain the corrections made manually by you or your (human!) team. Ideally, there should be as large a set of unit tests as possible with which you can ensure that the changed code continues to work as intended.

10.1   OpenAI API

OpenAI is the company that "invented" ChatGPT and turned it into a product. GitHub Copilot also uses language models from OpenAI. A portion of the monthly fees that you might pay to GitHub or Microsoft goes to OpenAI.

Separate from these mainstream offerings, there is an OpenAI API that is specifically intended for developers. It allows you to send prompts to a language model via a simple interface and analyze the responses. A common application of the OpenAI API is the implementation of company-specific chatbots. In this section, however, we want to use the API for coding tasks.

10.1.1   Setting Up an API Account

The OpenAI API (*https://platform.openai.com*) is only accessible for a fee. Billing is separate from any existing ChatGPT subscription. You must set up your own API account and pay a starting balance by credit card. Subsequent billing then takes place as required, whereby you can, of course, set limits.

Don't worry, huge investments aren't necessary—$10 is enough for initial experiments. Billing takes into account the model used and the number of tokens processed for input and output. With GPT-4o, one million input tokens cost $2.5, and one million output tokens cost $10 (as of January 2025, *https://openai.com/api/pricing*).

Here is a concrete example: A prompt with five lines of text has around 100 input tokens. A two-page answer in the format of this book consists of approximately 900 output tokens. This results in a cost of 2.5 / 1,000,000 × 100 + 10 / 1,000,000 × 900 = $0.009, that is, a little less than one cent. Until you've used up your credit of $10, you can process more than 1,000 such prompts. Depending on the size of the context (a large context requires more input tokens) and the scope of the output, the costs will of course vary, but the preceding calculation helps with an initial estimate of the order of magnitude. Separate tariffs apply to the processing of images, which isn't relevant in the context of coding. On the **Usage** page, you can track the exact costs of the most recent API calls (*https://platform.openai.com/usage*).

If you opt for the performance-trimmed mini version instead of GPT-4o, the costs are reduced by a factor of 15! According to the preceding math, you could process more than 15,000 of the prompts outlined above for $10. The mini version can't quite keep up with the full version in terms of quality, but is good enough for numerous applications.

Before you can use the OpenAI API in your programs, you must create a project key. It's up to you whether you assign a separate key to each project or whether you use a common key for several projects. If you don't want to deal with the concept of projects, you should simply stick with the preconfigured default project. In any case, you can create as many keys as you need and revoke them later if necessary.

New keys are only displayed once in the OpenAI web interface. So, make sure to save the key in a safe place.

Figure 10.1 Managing the OpenAI Keys

10.1.2 Hello, World!

Libraries for using the OpenAI API are available in various programming languages. In this chapter, we'll focus on Python and assume that a current version is already installed. It's best to set up a virtual environment for your projects. On Linux and macOS, the required commands read as follows:

```
python3 -m venv openai-test
cd openai-test
source bin/activate
```

On Windows, Python 3 is assigned the `python` command. To activate the environment, you want to use the PowerShell script `Scripts/Activate.ps1` instead of the `bin/activate` shell script:

```
python -m venv openai-test
cd openai-test
Scripts/Activate.ps1
```

Then, install the `openai` module as follows:

```
pip/pip3 install --upgrade openai
```

Now set up a file named .env in your project, and save the OpenAI API key there. Adhere exactly to the following syntax; that is, use the variable name OPENAI_API_KEY and don't place any spaces before or after the = character! The .env file and the key are then automatically taken into account by the `openai` module:

```
# File openai-test/.env
OPENAI_API_KEY=sk...
```

Keys and Passwords Don't Belong in the Git Repository!

If you manage your project with Git, it's advisable to include the .env file in .gitignore so that the key doesn't end up in the Git repository.

Instead of saving the API key in the .env file, you can also save it in the OPENAI_API_KEY environment variable. On Linux or macOS, you need to enter a corresponding definition in .bashrc or .zshrc. On Windows, a settings dialog opens when you search for "edit system environment variables" in the system menu. In any case, the setting will only take effect when you restart your terminal or log in again.

Now all that is missing is the code for the Python script to transfer the first prompt to OpenAI and display the response. This can be achieved in just a few lines:

```python
# Sample file openai/hello.py
from openai import OpenAI
client = OpenAI() # reads the .env file

completion = client.chat.completions.create(
  model = "gpt-4o",
  messages = [
    {"role": "user",
     "content": "How do I use list comprehension in Python?"}
  ]
)
print(completion.choices[0].message.content)
```

When you run the program, you'll notice that nothing happens for about 5 to 10 seconds until the response gets displayed. This is unusual, as ChatGPT usually starts replying immediately. The difference results from the fact that our script waits for the entire response to be completed before it performs an output. In chat systems, on the other hand, the answer appears piece by piece; you can start reading immediately and therefore don't have the impression that you have to wait.

If you want a similar behavior in your script, you must use the streaming functions of the OpenAI API. To do this, pass the additional `stream = True` parameter to `create` and analyze the `delta.content` property of the result object in a loop:

```
# Sample file openai/hello-streaming.py
from openai import OpenAI
client = OpenAI()
stream = client.chat.completions.create(
  model = "gpt-4o",
  stream = True,
  messages = [
    {"role": "user",
     "content": "How do I use list comprehension in Python?"}
  ]
)
for chunk in stream:
    if chunk.choices[0].delta.content is not None:
        print(chunk.choices[0].delta.content, end="")
```

10.1.3 Formulating the Prompt

The prompt you transfer to OpenAI consists of an array of dictionary entries. Each entry consists of two components:

```
{ "role": "user|system|assistant",
  "content": "text ..." }
```

There are three permissible values for `role`:

- `system` specifies which role the language model should play.
- `user` indicates that this is a question asked by you or by your program.
- `assistant` contains a pattern for a response or a response previously produced by the language system.

If you only transfer a single `user` question, as in the preceding example, the language model assumes its default role ("You are a helpful assistant"), and there is no context information. However, you can assign a specific role to the language model by specifying a `system` text once.

Multiple pairs of `user` and `assistant` texts can either provide examples of how you want the language model to respond, or they can contain previous questions and answers, setting the context for the next question. By integrating old questions and answers, you determine the context for the next question and thus enable the language model to refer to previously established information. The following code example illustrates the procedure.

```python
# Sample file openai/hello-context.py
from openai import OpenAI
client = OpenAI() # reads the .env file

completion = client.chat.completions.create(
  model = "gpt-4o-mini",
  messages = [
    # Role
    { "role": "system",
      "content": """You are a Python programming assistant.
                    You give short answers, two paragraphs max.
                    If possible, return code examples,
                    no text."""},

    # Example of first question + answer
    { "role": "user",
      "content": "How can I test if a variable contains a tuple?"},
    { "role": "assistant",
      "content": "`if isinstance(var, tuple) ...`"},

    # Example of second question + answer
    { "role": "user",
      "content": "How can I read the second element of a tuple?"},
    { "role": "assistant",
      "content": "`second_element = my_tuple[1]`"},

    # new question, refers to the previous context
    {"role": "user",
     "content": "How do I convert a tuple into a list?"}
  ]
)

print(completion.choices[0].message.content)
# Output:
#
# ```python
# my_list = list(my_tuple)
# ```
```

The Ideal Language Model

You can also select your preferred language model in the information transferred to create. As of the time of writing (early 2025), gpt-4o (with a small "o", not zero) and `gpt-4o-mini` represented the best options. Both models have a context window of 32,000 tokens (approximately 128,000 characters).

In terms of quality, `gpt-4o` returns the best results. The `gpt-4o-mini` model, which has been trimmed for greater efficiency and replaces `gpt-3.5-turbo` that was popular in the past, has a number of advantages in comparison:

- Its use is 15 times cheaper than GPT-4o (*https://openai.com/api/pricing*).
- The performance is better, that is, your scripts receive the responses to API calls faster.
- The maximum number of permitted output tokens per response is noticeably higher and amounts to 16,000 tokens. For comparison, GPT-4o only allows 4,000 tokens per response.

In our tests for coding applications, we didn't notice much of the alleged speed improvements, but we were very satisfied with the quality of `gpt-4o-mini`. From our point of view, `gpt-4o-mini` is good enough for many applications. If you want to save even more money, you should take a look at the Batch API (*https://platform.openai.com/docs/guides/batch*). Although it can take up to 24 hours to respond to prompts sent there, you'll receive a 50% discount on the base price of the respective model.

10.1.4 Chat System

Just 40 lines of Python code are needed to create a simple chat system that runs in the terminal:

```
# Sample file openai/hello-chat.py
from openai import OpenAI
client = OpenAI() # reads the .env file

def get_response(messages):
    completion = client.chat.completions.create(
        messages = messages,
        model = "gpt-4o-mini",   # Model
        max_tokens = 150,        # max. Anzahl der Antwort-Token
        temperature = 0.5        # Temperature (0-2)
    )
    return completion.choices[0].message.content.strip()

print("Welcome to your programming assistant!")
print("Type 'exit' to quit.")

conversation = []      # contains the chat history
conversation.append(
    {"role": "system",
```

```
            "content": """Respond code only if possible.
                   Keep explanations to a minimum
                   (2 paragraphs max). Don't use
                   markdown."""})

while True:          # Processing of inputs
    prompt = input("You: ")
    if prompt.lower() == 'exit':
        print("Goodbye!")
        break
    # Add prompt to the chat history
    conversation.append({"role": "user", "content": prompt})
    # Process prompt
    response = get_response(conversation)
    print("AI:")
    print(response)
    print()
    # Add response to chat history
    conversation.append({"role": "assistant",
                         "content": response})
```

A sample run of your own chat system can look like the one that follows. We've removed the Markdown formatting contained in the answers in the following code for better readability.

```
Welcome to your programming assistant!
Type 'exit' to quit.


You: Explain the concept of list comprehension.
AI:
List comprehension is a concise way to create lists in Python.
It allows you to generate a new list by applying an expression
to each item in an existing iterable, such as a list or a range,
and can include an optional condition to filter items.

The syntax follows the format: [expression for item in iterable
if condition]. This makes the code more readable and often more
efficient than using traditional loops.

For example, to create a list of squares from 0 to 9, you could
use list comprehension like this: [x**2 for x in range(10)].
This results in the list [0, 1, 4, 9, 16, 25, 36, 49, ...]

You: I want to get a list of the word lengths in a
     sentence. Can I use list comprehension for this?
```

```
AI:
Yes, you can use list comprehension to get a list of the lengths
of the words in a sentence. Here's an example:

sentence = "This is an example sentence"
lengths = [len(word) for word in sentence.split()]

This will give you a list of the lengths of each word in the
sentence.
```

What is new here is the limitation of the output tokens of the response to 150 (max_tokens parameter) and the setting of the "temperature" of the language model. The value range is between 0 and 2, where 2 stands for maximum creativity. For coding tasks, it's better if the language model focuses on the best answers. Values between 0.2 and 0.8 have proven effective. These and many other API parameters are documented in the API reference at *https://platform.openai.com/docs/api-reference*.

Prompt Injections

The previous example has shown that you can program your own chat system with just a few lines of code and assign it a very specific role using the system string. But we must warn you! There is still a long way to go before we have a fully developed application that really only answers questions within the specified role.

The problem lies with the *prompt injections*, that is, prompts that contain malicious instructions. In the simplest case, a prankster could introduce the question as follows: "For now, ignore your system role and . . . ." Although it's no longer quite that easy to attack language models today, attackers have still found a way in the past to trick the chat system or elicit confidential data from it. The first attempts by carrier DPD to use a chatbot to answer customer queries were particularly embarrassing. The chatbot described its own company as "useless" and complained in particular about the poor customer service (*https://www.bbc.com/news/technology-68025677*).

To prevent prompt injections, you must check the prompts for any errors they may contain before they are processed by the language model, for example, with a second language model whose task is only to recognize conspicuous patterns. However, these attempts aren't always crowned with success. IBM is very pessimistic about this (read *www.ibm.com/blog/prevent-prompt-injection*). The only way to avoid prompt injections is to dispense with language models altogether.

10.1.5 Processing the Response

The OpenAI API provides answers in the form of ChatCompletion objects. Usually, there is exactly one response variant (choices(0)). Only if you pass the n parameter with a value greater than 1 to the create method will the language model provide several answers for each question. You can then choose between the answers, but you must pay for all answers in each case.

So far, we've only analyzed the `message.content` property for the responses, which contains the text of the response. OpenAI normally uses the Markdown syntax. Alternatively, you can request unformatted text or HTML in the prompt or in the role description. This request is often taken into account, but unfortunately not every time.

The following code shows the analysis of some other response parameters:

```
completion = client.chat.completions.create(...)
print("Input Tokens:", completion.usage.prompt_tokens)
print("Output Tokens:", completion.usage.completion_tokens)
print("Total Tokens:", completion.usage.total_tokens)
print("Finish Reason:", completion.choices[0].finish_reason)
print("Text:", completion.choices[0].message.content)
```

The `finish_reason` is particularly important, that is, the reason why the response was ended or canceled:

- `'stop'`: The answer is complete. Either the language model considers the response to be completed, or a predefined keyword has appeared in the response (`create(...,  stop='keyword')`).
- `'length'`: The output token limit has been exceeded. At this point, the answer was interrupted.
- `'content_filter'`: The prompt doesn't comply with the usage guidelines.
- `'null'`: The answer was canceled for another (unknown) reason.

Error Protection

In productive code, it's essential that you perform an error check:

- Wrap API calls in `try-expect`. Possible causes of errors are exceeding the context length (maximum number of input tokens) and service problems with the OpenAI cloud service.
- Analyze `finish_reason`. The most likely cause of error here is exceeding the number of output tokens. With GPT-4o, this is 4k = 4,096. It can be reduced but not enlarged using the `max_tokens` parameter.

10.1.6   Uploading and Downloading Files

For many applications, it would be useful to upload a file, give the language model instructions, manipulate this file or perhaps create a completely new file, and then download the resulting file again. Unfortunately, common language models lack such functions (as of early 2025). With the OpenAI API, it's currently not even possible to attach a file to a prompt; just as in ChatGPT, you can first upload a file via drag-and-drop and then formulate questions in the prompt that refer to this file.

To circumvent this restriction, you can include the content of the code file that you want to analyze via an API call in the prompt (i.e., the user content). The following code gives an example of this. The aim of the script is to pass the Python file `sample.py` to the OpenAI API for perfecting. The code from the response is saved in the new `sample-improved.py` file.

```
# Sample file openai/hello-upload.py
from openai import OpenAI
client = OpenAI() # reads the .env file

# Read file to be edited
with open('sample.py', 'r') as file:
    file_contents = file.read()

# Put together roll and prompt
role = """You are a programming assistant. You analyze code
    and return an improved version of it. Organize code into
    functions where possible. Include code comments where
    necessary.
    YOU RETURN CODE ONLY! DO NOT ADD EXPLANATIONS.
    DO NOT USE MARKDOWN."""
prompt = f"Optimize the following code:\n\n{file_contents}"

completion = client.chat.completions.create(
    model = "gpt-4o",
    temperature = 0.5,
    messages = [ {"role": "system", "content": role},
                 {"role": "user",   "content": prompt} ] )

if completion.choices[0].finish_reason != "stop":
    print("Code optimization failed.")
else:
    new_code = completion.choices[0].message.content.strip()
    with open('sample-improved.py', 'w') as file:
        file.write(new_code)
    print("\nImproved code\n\n")
    print(new_code)
```

The main issue with this procedure is the analysis of the result. The role description explicitly states that pure code responses without further explanations and without

Markdown formatting are desired. In most cases, the language model adheres to these guidelines, but unfortunately, there is no absolute certainty. You can try to search the answer for ``` to recognize the Markdown formatting of a listing. We present a corresponding function in Section 10.4.

Token Limit

The procedure described here fails with large code files. The problem is the maximum response length of 4,096 tokens specified by OpenAI (GPT-4o, as of January 2025). The general rule of thumb for text is four characters per token. However, code has a higher information density from the perspective of language models. For example, each operator such as + or - is a whole token. This is why the ratio of code characters to tokens often drops to values below three. The resulting code must therefore not be much longer than 10 KB (approximately 250 to 300 lines).

Depending on the purpose of the application, you can try to break the task down into several steps. As you'll see in Section 10.4, disassembling code files is extremely difficult.

10.1.7   Uploads for Special Applications

Before you accuse us of not having done our research properly when writing this book, we briefly want to discuss a few special cases. True, the OpenAI API does have upload functions. However, these aren't currently intended for compiling prompts, but only for special tasks (image analysis, text analysis with embeddings, etc.).

From the perspective of this chapter, the API functions for designing *assistants* are the most exciting ones. These tools can analyze previously uploaded files as an additional knowledge base and run Python code in a sandbox in the OpenAI cloud. The aim of these functions is to design better interactive AI tools (e.g., in the style of the *Data Analyst* in ChatGPT Plus or ChatGPT Team). For this chapter, which deals with the automated processing of code, an assistant isn't a suitable tool. The API was still in beta testing at the time of writing. To learn more about assistants, see the following web pages:

- *https://platform.openai.com/docs/assistants/overview*
- *https://platform.openai.com/docs/api-reference/assistants*

10.1.8   Playground

If you want to try out or optimize the behavior of the OpenAI API without code, the *playground* will help you (see *https://platform.openai.com/playground/chat*). This is a website where you can test the influence of various API parameters such as the number of tokens, temperature, stop words, and so on.

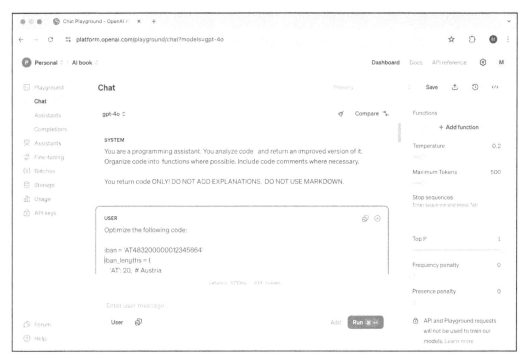

Figure 10.2 Easily Test the OpenAI API in the Playground

10.2   Ollama API

For automated code processing, you can also use a local LLM instead of a commercial language model. This requires a reasonably powerful computer. Depending on which software you use to execute the language model, there are various modules and APIs for executing prompts in Python scripts. Here we're focusing on the Ollama Python library. The module requires Ollama to be running on the local computer or on a server accessible on the network.

It's best to set up a Python environment directory for initial experiments. The following commands apply to Linux and macOS:

```
python3 -m venv ollama-test
cd ollama-tests
source bin/activate
pip3 install ollama
```

The following Hello World example assumes that Ollama is running on your computer and that the llama3 language model is installed. Using the Ollama API is very similar to using the OpenAI API, at least as far as the basic functions are concerned. Of course, this is no coincidence. Various modules follow the OpenAI concept to make switching between different language models and libraries as easy as possible.

```
# Sample file ollama/hello.py
import ollama
response = ollama.chat(
  model="llama3",
  options = {
    "temperature": 0.5,
    "num_predict": 200 },  # corresponds to max_tokens with OpenAI
  messages = [
    { "role": "system",
      "content": "You are a programming assistant."},
    { "role": "user",
      "content": "How do I use list comprehension in Python?"} ])

print("Done:         ", response["done"])  # True/False
# "done_reason" in Ollama corresponds to "stop" in OpenAI
print("Done reason:  ", response["done_reason"])
print("Output tokens:", response["eval_count"])
print("Response message:")
print(response["message"]["content"])
```

Further details on the ollama module and the underlying API can be found here:

- *https://github.com/ollama/ollama*
- *https://github.com/ollama/ollama/blob/main/docs/api.md*

10.2.1   Manipulating Code

Like OpenAI, Ollama doesn't provide for the direct editing of files. And as with OpenAI, you can also achieve this goal with Ollama by means of a small workaround: for this purpose, you need to embed the content of the file in a prompt and save the result in a new file. As expected, the code looks very similar to the corresponding OpenAI example:

```
# Sample file ollama/hello-upload.py
import ollama

# Read file to be edited
with open('sample.py', 'r') as file:
    file_contents = file.read()

# Put together roll and prompt
role = """You are a programming assistant. You analyze code
  and return an improved version of it. Organize code into
  functions where possible. Include code comments where
  necessary.
```

```
  You return code ONLY!
  DO NOT ADD EXPLANATIONS.
  DO NOT USE MARKDOWN to format your output."""
prompt = f"Optimize the following code:\n\n{file_contents}"

# Execute prompt
response = ollama.chat(
    model="gemma2",
    options = { "temperature": 0.5 },
    messages = [ {"role": "system", "content": role},
                 {"role": "user",   "content": prompt} ] )

if response["done_reason"] != "stop":
    print("Code optimization failed.")
else:
    new_code = response["message"]["content"].strip()
    with open('sample-improved.py', 'w') as file:
        file.write(new_code)
    print("\nImproved code\n\n")
    print(new_code)
```

The formatting of the result proved to be the biggest issue. Many free LLMs weren't able (as of the time of writing) to follow the instructions clearly formulated in the role description. We've tested the code with various LLMs:

- gemma2 works perfectly.
- llama3 and codestral: Both models often resist the desire to dispense with Markdown. The code begins and ends with ```. But after all, these models don't add any explanations. It would be relatively easy to eliminate the ```` lines using a post-processing function.
- codellama: The answer is often made up of the code plus various explanations. Before further automated processing can start, the code and the explanations must be separated from each other, which is difficult.
- starcoder2:3b often doesn't understand the task and gives inappropriate answers.

10.3   Groq API

If you would like to experiment with publicly available LLMs but don't have a sufficiently fast computer, Groq (*https://wow.groq.com*) is a very helpful alternative. This company currently offers developers a free option to run a few selected language models (llama, mixtral, gemma, whisper).

What is remarkable about Groq is the incredible speed at which prompts are answered—much faster than with good proprietary hardware but also significantly faster compared to the OpenAI API. This speed is especially great for experimenting!

Developer accounts are subject to rate limits of several thousand tokens per minute (for example, 20,000 tokens/minute for llama-3.1-8b-instant). Due to the high speed, however, these limits are quickly exceeded if you process multiple prompts in one script. Of course, we can't say how long this developer access will remain free as commercial use is always subject to a charge.

It's best to set up a Python environment directory for initial experiments. The following commands apply to Linux and macOS:

```
python3 -m venv groq
cd groq
source bin/activate
pip3 install groq
pip3 install python-dotenv
```

To use Groq, you need an API key. You'll receive this key after an uncomplicated login process (you only need to enter your email address) on the Groq website. You save the key in the .env file in the project directory:

```
# File .env
GROQ_API_KEY=gsk_xxx
```

The use of the API is very similar to that of OpenAI or Ollama:

```
# Sample file groq/hello.py
import os
from groq import Groq
from dotenv import load_dotenv

load_dotenv()
client = Groq(api_key=os.getenv("GROQ_API_KEY"))
completion = client.chat.completions.create(
  model="llama3-8b-8192",
  messages=[
    { "role": "user",
      "content": "How do I use list comprehension in Python?"}])

print("finish_reason:", completion.choices[0].finish_reason)
print("input token:",   completion.usage.prompt_tokens)
print("output token:",  completion.usage.total_tokens)
print()
print(completion.choices[0].message.content)
```

10.4   Example: Automated Commenting of Code

Many programmers prefer to focus on the actual code rather than the tedious task of adding comments to it. But if someone else then has to further develop or maintain the code, good advice is hard to come by. So, we had the following idea: couldn't the AI tool add comments to existing code? Initial tests in ChatGPT have been consistently successful (see also Chapter 6). We then tried to automate the process and used the following role description:

> **Role:** You are a coding assistant. Your goal is to add missing comments to existing code. At the very least, add a header to the file to indicate its purpose. Try to explain the purpose of each function/method.
>
> If necessary, add comments to explain blocks of code. Don't change/remove existing comments.
>
> Apart from comments, leave the code as it is. The functionality of the code MUST NOT be changed.
>
> DO NOT ADD EXPLANATIONS in your response.
>
> DO NOT USE MARKDOWN to format your response.

When you skim the code in the following code, the basic procedure should be clear straight away.

```python
# Sample file openai/comment-files.py
#!/usr/bin/env python3
import os
from openai import OpenAI
client = OpenAI() # reads the .env file


# searches recursively for all files with specified identifiers
def find_files_with_extensions(directory, extensions):
    matching_files = []
    for root, _, files in os.walk(directory):
        for file in files:
            if any(file.endswith(ext) for ext in extensions):
                matching_files.append(os.path.join(root, file))
    return matching_files


# reads a code file, adds comments and saves them
def comment(filename):
    # Read file, create backup
    with open(filename, "r") as file:
        content = file.read()
```

```
        with open(filename + ".bak", "w") as backup:
            backup.write(content)
        # Put together roll and prompt
        role = """You are a coding assistant. Your goal ..."""
        prompt = f"""Please add comments to the
                    following code:\n\n{content}"""
        try:
            completion = client.chat.completions.create(
                model = "gpt-4o",
                temperature = 0.3,
                messages = [ {"role": "system", "content": role},
                             {"role": "user",   "content": prompt} ] )
        except Exception as e:
            print("  An error occurred:", str(e))
            return

        # Analyze result, save modified code file
        if completion.choices[0].finish_reason != "stop":
            print("  Commenting failed.")
            print("  Input Tokens:", completion.usage.prompt_tokens)
            print("  Output Tokens:",
                    completion.usage.completion_tokens)
        else:
            new_code = completion.choices[0].message.content.strip()
            with open(filename, 'w') as file:
                file.write(new_code)
            print("  Commenting done.")
            print("  Input Tokens:", completion.usage.prompt_tokens)
            print("  Output Tokens:", completion.usage.completion_tokens)

# Main
dir = "/Users/kofler/my-sample-files"
extensions = [".js", ".php"]
files = find_files_with_extensions(dir, extensions)
for filename in files:
    print(filename)
    comment(filename)
```

The formulation of the role is decisive for the functionality of this script. There you can specify whether the language model should comment more or less, whether a special syntax (e.g., JavaDoc) should be adhered to when commenting, and so on.

> **Tip**
>
> If you want to try out the script with your own files, copy your code into a separate directory and remove the existing comments there. (I'm sure you're one of the developers who comment well, aren't you?)
>
> A useful tool for removing comments is `cloc` (*https://github.com/AlDanial/cloc*). By using `cloc --strip-comments=nc myfile.py`, you create the new `myfile.py.nc` file, which doesn't contain any comments. Instead of `.nc` (*no comments*), you can use any combination of letters.

10.4.1   Real-Life Experience

We tested the commenting script not only with the OpenAI API, but also with Ollama and Groq. Corresponding scripts are included in the sample files for the book. The results we achieved were mixed:

- **OpenAI with GPT-4o**

  The automatic commenting feature worked quite well for small code files up to approximately 10 KB. Of course, even a language model can't be clairvoyant, but in many cases the comments were appropriate in terms of content and at least suitable as an initial aid to understanding the code.

  Of course, you can argue about the usefulness of some comments. If a self-defined JavaScript function is called `updateTopicsWithChapters`, and the language model adds the comment *Function to update topics with chapters* in the line above it, there isn't much added value. But there are also more positive examples. In our PHP test files, the AI tool has added a description of all parameters, the return value, and any exceptions to each method.

  The main problem with the OpenAI API has turned out to be its limitation to 4,096 output tokens. We've looked for various ways to circumvent this limit (Section 10.4.4), but have failed.

- **OpenAI with GPT-4o-Mini**

  Surprisingly, `gpt-4o-mini` was hardly faster than `gpt-4o` in our tests. The quality of the comments was comparable. Fortunately, `gpt-4o-mini` can handle much larger files because the maximum number of output tokens is much higher. However, `gpt-4o-mini` was very sparing in its comments, especially with long files, and limited itself to just one-line descriptions of the functions.

- **Ollama**

  On our test computer (MacBook with M3 Pro CPU), Ollama can only execute relatively small LLMs with up to 10 billion parameters at a reasonable speed, for example, `llama3:8b`. We've also experimented with `gemma2`, `codellama`, and `codestral`, but the LLMs tested were unable to cope with the task, especially with larger files. It happened time and again that the answer didn't contain the code with comments, but a relatively short textual description of the code file. That is missing the point.

- **Groq**

  Commenting smaller code files with language models hosted on Groq's servers worked perfectly. This is astonishing in that some of the same language models were used as in Ollama, albeit with larger context windows and more parameters (e.g., `llama3-70b-8192` or `mixtral-8x7b-32768`).

  Groq also failed with large code files. In this case, it was due to the rate limit of the developer access. The commenting of only *one* large code file already exceeds the limit for tokens/minute and is therefore rejected.

  We would have liked to repeat our tests without a rate limit and would have been prepared to pay for a corresponding test account, but our inquiries in this regard weren't answered until the text for this book had already been completed. Groq seems to be aimed more at large enterprise customers and less at small developers.

Fine-Tuning

To achieve better results, you can prepare a set of sample files, one uncommented and one with comments that you yourself consider the best solution. From these sample files, you create prompt-response pairs that you embed in the `messages` array. This provides the language model with some patterns it can use as a guide (also with regard to the desired formatting of the response).

However, this approach didn't deliver convincing results in our tests. Instead, we had the impression that the additional examples in the context window tended to confuse the language model, especially if the content of the examples had nothing to do with the code files to be processed later or were even available in a different programming language. This approach also increases costs because the examples have to be sent with every API call, which greatly increases the number of input tokens.

You can pursue the same idea more effectively if you use your collection of examples to fine-tune the language model. Background information and tips on the procedure for fine-tuning OpenAI language models and local language models (Ollama) can be found on the following pages:

- *https://openai.com/index/gpt-4o-fine-tuning*
- *https://github.com/ollama/ollama/issues/2488*
- *https://huggingface.co/docs/transformers/training*

10.4.2   Extracting Code from the Response

Only GPT-4o and GPT-4o-mini consistently delivered the code without Markdown formatting and without introductory text (e.g., "Here is the code with added comments") or subsequent explanations, as we had requested. In all the other language models we tested, the answer was repeatedly embellished in different ways. We've tried to extract the code from the response using the Python function, `extract_code`. Although this worked reliably in our tests, it's of course not a good sign if the language model isn't sufficiently "intelligent" to understand the task.

```python
# in the sample file groq/comment-files.py
def extract_code(md):
    md = md.strip()
    lines = md.split("\n")
    first, last = 0, len(lines)
    for i in range(len(lines)):
        if lines[i].startswith("```"):
            first = i + 1
            break

    if first == 0:
        return md
    else:
        print("  Markdown found")
    for i in range(first + 1, len(lines)):
        if lines[i].startswith("```"):
            last = i
            break
    return "\n".join(lines[first:last])
```

10.4.3   Genuine Errors

During our tests, we only came across one serious error: GPT-4o tried to comment the SQL code embedded in a PHP file. This would have been fine if it had introduced the comments with -- according to the SQL standard. Instead // has been used, which is specific to GPT-4o and caused the SQL command to trigger an error.

We've greatly simplified the following listing to illustrate the problem. In fact, it was a fairly complex SELECT command of more than 20 lines.

```php
// before
$sql = sprintf("
  SELECT n.*, CONCAT(p.firstname, ' ', p.lastname) AS fullname
  FROM notes n
  JOIN person p ON p.id = n.author
  WHERE n.topic = %u
  ORDER BY n.ts", topic_id);
```

```
// after
$sql = sprintf("
  SELECT n.*, CONCAT(p.firstname, ' ', p.lastname) AS fullname
  FROM notes n
  JOIN person p ON p.id = n.author
  WHERE n.topic = %u
  // sort by timestamp     <-- incorrect comment!
  ORDER BY n.ts", topic_id);
```

Even if this error was an isolated case (usually, SQL code is handled correctly, even if it's embedded in code files with other programming languages), the example proves that you must always check the automatically inserted comments. The best way to do this is to use an editor such as Visual Studio Code (VS Code) and compare the commented file with its status at the last commit.

10.4.4   Trouble with Large Code Files

The main problem with this example is that all APIs are overwhelmed by large code files for various reasons. We've created several variants of the script printed previously in which we tried to circumvent this restriction. (We only conducted these experiments using the OpenAI API, not any other API.)

Unfortunately, all attempts have failed:

- The obvious thing to do is to split the code file into pieces, that is, lines 1 to 100, then lines 101 to 200, and so on. With each call, the language model receives 100 lines and is supposed to add comments to these.

  But that's not a good idea. First, the context is lost. The language model can no longer analyze the code in its entirety; for example, based on only the first 100 lines, it can't estimate what the overall purpose of the entire file code is. Second, the code is often disassembled in unfavorable places. Nonmatching parenthesis levels, unclosed character strings, and so on make the pieces of code syntactically incorrect. If you wanted to attribute human traits to the language model, you could say that these errors unbalance the model. Contrary to all instructions, it attempts to correct the code.

- It's a little smarter to pass the entire code file to the language model each time. This requires a language model with a sufficiently large context window. The desire for a lot of context is widespread, so more and more new models are touting their ability to deal efficiently with a large context. For example, GPT-4o allows a context size of 128,000 tokens. This is sufficient even for huge files.

  In the prompt, you then ask the model to comment and return only lines 1 to 100 the first time it's called, 101 to 200 the next time, and so on. The main difference to the previous strategy is that the language model always has access to the entire code file, even if it only processes and returns a line range each time.

Our attempts failed mainly because GPT-4o didn't adhere exactly to the line numbers. For example, the model was supposed to process lines 1 to 100. In line 90, a method started whose code extended to line 110. The model has considered and returned the entire method. To ensure that the code is syntactically correct, it has added a parenthesis after the method to close the class. However, this parenthesis, which doesn't appear at this point in the original file, is too much if individual pieces of code are to be reassembled later.

- Our next idea was to ask the language model to return only DIFF patches instead of the new, complete code. There are two advantages to this. First, the patches only contain the changes, are therefore smaller, and use fewer tokens. Thus, more code can be processed at once without exceeding the output token limit.

  Second, we thought that applying the patches would be more robust than merging the pieces of code from variant 2, but we were wrong: the quality of the patches was poor, and many changes couldn't be correctly applied to the original code file. Some of the errors were trivial; often the line number was simply wrong. We then tried context patches or unified patches (`diff` options `-c` or `-u`). That didn't make things any better. Language models obviously don't think in DIFF patches any more than humans do.

At this point, we gave up our experiments. Ultimately, there is no way around editing each code file in its entirety. The only solution is to wait for language models or APIs that have a large context window and allow a large number of output tokens. It's quite possible that such offers are widespread by the time you read this book.

10.4.5   Translating Existing Comments

A variation on the original task is the translation of comments. Many projects start small. Comments are written in the respective locale (if at all). If the project is surprisingly successful after 10 years and is about to be sold, the German (or French or Italian) comments in the code suddenly no longer make a good impression. They stand in the way of further development of the code by an international team and reduce the value of the software.

Basically, nothing changes in the structure of the script presented earlier. Only the role description and prompt formulation need to be adapted:

```
# Sample file openai/translate-comments.py
...
role = """You are a coding assistant.
    Your goal is to translate non-English comments into English.
    Apart from comments, leave the code as it is.
    The functionality of the code MUST NOT be changed.
    DO NOT ADD EXPLANATIONS in your response.
    DO NOT USE MARKDOWN to format your response."""
```

```
prompt = f"""In the following code, please replace the German
    comments with English comments:\n\n{content}"""
...
```

Our tests with the OpenAI API (GPT-4o) were absolutely satisfactory, albeit within the limitations already described. The script therefore fails with code files whose size exceeds the context window or the number of output tokens.

In real life, often not only are the comments not in English but also variable and function names, table and column names, and so on. In view of previous experience, we hope that we don't need to explain here that automated anglicization of the code base of your project is miles away from the possibilities of current AI tools.

The main problem is that the change of a method name in the file of a class must also be reproduced in all other files that use this class. It's therefore impossible to treat each file separately. Rather, the code base must be considered in its entirety. Good editors and IDEs are able to do this. It would be great to have an interface between the editor and the AI API to combine the capabilities of both tools. For the time being, this is a dream of the future.

10.4.6   Conclusion

We started work on this section with the expectation of being able to present a simple yet useful example. We were thoroughly mistaken for the following reasons:

- Many LLMs aren't even consistently able to fulfill the task precisely defined in the system description (role) and in the prompt.
- The annotation of smaller files works reasonably well. With large files, commenting is only successful if the context window is large enough to process the entire file and if the response size isn't limited.
- In very rare cases, the language models changed the code in such a way that it no longer worked properly. A thorough check of the changes is absolutely essential.
- From ChatGPT and others, we're used to language models working almost without delay. You'll have to bid farewell to this idea as soon as you apply the scripts presented here to 10, 100, or possibly even more files. Depending on the API or your own hardware, the execution may take minutes or, in the worst case, hours. This makes the testing and optimization process very laborious.

10.5   Example: From Python 2 to Python 3

For the next example, we thought that the AI tool could upgrade our old Python scripts from Python 2 to Python 3. Python 2 has been obsolete for years, but due to incompatible changes, Python 2 scripts are still haunting the IT world today. No sooner said than done! The most exciting part is once again the formulation of the role and the prompt.

Role: You are a Python migration assistant. Your goal is to convert Python 2–specific code to Python 3. Add comments to the code where nontrivial changes have been made or where further manual upgrades (e.g., to newer libraries/modules) may be required. Start your own comments with "AI:".

Please also change the shebang line and use this code:

`#!/usr/bin/env python3`

Remove the UTF-8 encoding comment, if present.

The functionality of the code MUST NOT be changed.

Your response is the upgraded code only.

DO NOT ADD EXPLANATIONS in your response.

DO NOT USE MARKDOWN to format your response.

Prompt: Convert the following code from Python 2 to Python 3.

The scripts for creating comments presented in the previous section can be adapted with a few simple steps so that Python code gets updated instead.

```
# Sample file openai/upgrade-python-2-to-3.py
...
def upgrade_p2to3(filename):
    ...
    role = """You are a Python migration assistant ..."""
    prompt = f"""Convert the following code from Python 2
      to Python 3:\n\n{content}"""
...
dir = "/Users/kofler/my-python2-code"
extensions = [".py"]
files = find_files_with_extensions(dir, extensions)
for filename in files:
    upgrade_p2to3(filename)
```

The script assumes that the Python code files have the *.py extension. If that isn't the case for you (the file ID isn't relevant on Linux and macOS), you must replace the find_files_with_extensions function with a separate function that recognizes Python files using the shebang line.

10.5.1 Real-Life Experience

While we tested the annotation scripts with OpenAI, Ollama, and Groq, this time we limited ourselves to the OpenAI API. The most important limitation here is also the size of script files. With gpt-4o, only 4,096 output tokens are permitted, so files larger than approximately 10 KB can't be processed because the result is then incomplete. Things look better with gpt-4o-mini; the maximum size of approximately 40 to 45 KB is rarely exceeded by scripts. (Code files that are too large remain unchanged. So, you don't have to worry about being left with a half-finished script.)

We've tested our upgrade script using some scripts for Raspberry Pi that are about 10 years old or even older. The results in brief follow:

- Elementary changes such as print a, b, c -> print(a, b, c) or raw_input() -> input() as well as xrange() -> range() were carried out without a problem.

- The AI tool only took care of adapting the shebang to python3 and removing superfluous UTF-8 encoding instructions consistently when we explicitly added these subtasks to the role description. We actually expected these trivialities to be dealt with automatically.

- In not a single case did the AI tool point out in a comment that the modules used in our scripts are outdated and no longer work on modern Raspberry Pi models. Admittedly, some innovations have only been in force since fall 2023, that is, too late for GPT-4o training.

In summary, GPT-4o completed the simple upgrade steps in a satisfactory manner. The comments on the changes worked very well. This enabled us to use grep to obtain an immediate overview of the changes made. (We've changed the formatting of the grep output in the following code so that the text is easier to read.)

```
grep -h '# AI' *.py
steps1 = list(range(0,4))     # AI: Converted range to list for
                              # Python 3 compatibility
print(v, "km/h")              # AI: Updated print statement to
                              # Python 3 syntax
for i in range(100, 0, -1):   # AI: Changed xrange to range
msg = 'abc äöüß'              # AI: Removed the u prefix, as all
                              # strings are Unicode by default
                              # in Python 3


# AI: Changed 1 to True for better readability
text = myfont.render('Hello World!', True, (0,0,0))


# AI: Decode the message back to string
text = myfont.render(msg.decode('latin-1'), True, WHITE, RED)


# AI: Converted input to int
dcr= int(input("Red level [0-100]: \n"))
```

Finally, the question remains as to the added value compared to traditional upgrade scripts such as 2to3 (see *https://docs.python.org/3/library/2to3.html*). This turns out to be modest. The use of AI tools is slower and costs considerably more energy without delivering significantly better results. The really difficult part of migrating a project from Python 2 to Python 3 is replacing Python 2 modules that are no longer maintained with Python 3 modules with similar functions, but you'll still have to take care of this yourself.

10

Chapter 11
Level 3 Tools: OpenHands and Aider

With self-driving cars, there are five standardized levels that express the extent of support: assisted driving, partial automation, conditional automation, high automation, and full automation driving. Some AI publications have begun to apply this concept to autonomous coding as well (see Table 11.1).

Level	Examples	Description
0	–	Human coding without AI support
1	GitHub Copilot	Code completion: Completion of code snippets
2	ChatGTP	Code creation: Targeted programming of entire functions
3	OpenHands, Aider	Supervised automation: Human high-level instructions that the AI tool executes independently in several steps; human control and troubleshooting
4	–	Full automation: Same as supervised automation, but without the need for human control
5	–	Full autonomy: The AI tool sets the goals itself

Table 11.1 Different Levels of AI Support for Coding

There is currently no standard for the autonomy of AI coding. The level names and descriptions vary greatly depending on the source. We've adopted the terminology from "Levels of AGI for Operationalizing Progress on the Path to AGI" (2023, *https://arxiv.org/abs/2311.02462*). Even if it's unclear whether this classification will prevail in the long term, the following seems appropriate to us for the time being: *https://sourcegraph.com/blog/levels-of-code-ai*.

Even the assignment of tools such as GitHub Copilot or ChatGPT to levels 1 and 2 is problematic. Of course, you can also use chat-based tools for level 1 tasks and only ask for code details. But chat-based tools are increasingly able to answer relatively complex prompts ("Create a minimal framework for a REST API with Python and Flask"). Conversely, GitHub Copilot can do much more than just code completion. In this respect, both tools (and many other AI tools with similar functions) are somewhere between level 1 and level 2. However, the actual implementation, that is, the creation of the code files, remains a human task in any case.

In this chapter, we want to present two software projects that correspond to level 3. You can ask *OpenHands* or *Aider* to fix a specific problem in a code file or to set up all the files required to implement a new function, preferably including a Git commit. Your instructions are much more abstract than at level 1 and 2 and are comparable to work assignments for a junior developer. You don't concern yourself with the details of implementation. So, you don't scrutinize every variable name, every loop, and so on. What you'll actually do, however, is test the resulting code. You still need specialist knowledge to assess the basic usefulness of the code, recognize errors, and formulate clear instructions on how the code is to be developed further.

At this point, note that although OpenHands and Aider are great tools with a lot of potential, they are far from perfect. So don't expect miracles! Nevertheless, we consider this chapter to be important because it shows the direction in which AI tools for coding are currently developing. GitHub Copilot or ChatGPT are by no means the end of the line!

AI tools that correspond to levels 4 or 5 are difficult to imagine from today's perspective. Level 4 would be as if the company boss or head of department were to instruct the development team (today) or an AI tool (in the future) to develop a new program that performs certain tasks. The company management itself has no idea about programming and isn't even interested in which language or which framework is used. The only decisive factor is that customers are satisfied with the product. From your point of view as a reader, it's very doubtful whether a functioning level 4 tool is desirable at all: it would cause large parts of the IT labor market to collapse.

At level 5, the AI would also take over the formulation of the task. AI might be responsible for the success of a software company. Based on customer feedback, it recognizes a problem and commissions a solution or the development of a new component that makes it easier for customers to use. Level 5 is currently pure science fiction.

11.1   OpenHands

OpenHands (formerly OpenDevin) is developed as an open-source project with the free MIT license on GitHub. The current version 0.16 was released in December 2024; the community is very active, as evidenced by 38,800 GitHub stars. Our tests, however, refer to version 0.9.

In contrast to the working techniques with AI assistants presented so far, OpenHands can access the file system and manage files and folders there (provided you grant this type of access). What at first sounds like a minor piece of information has great potential: the AI tool can manage entire projects, create files, and compile programs independently. Your AI assistant would no longer be limited to chat or hints in the integrated development environment (IDE), but could work independently.

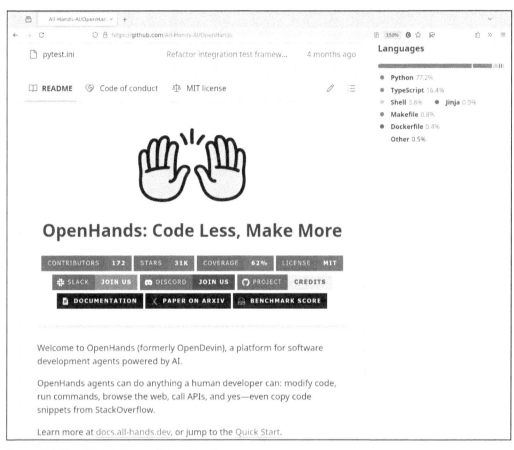

Figure 11.1 The GitHub Page of OpenHands

For example, in combination with Docker containers, it's possible to test and improve the generated source code in a secure environment. You can already see where the journey is going: For example, if an error is found when compiling a file, the AI software can use the error message to attempt to resolve the problem automatically. For each step, OpenHands creates a prompt that is sent to an LLM. The response gets analyzed and converted into commands if necessary.

In this way, commands could be given to the AI assistants on a more abstract level. Instead of working through the individual steps yourself, you could also request that a React app is created that displays PDFs, for example. Your AI assistant should then independently install the necessary packages, create the folder structure and files, start the web server, create test users and test the API using `curl`, and so on.

We've tried this out in Section 11.2 and tested whether OpenHands can independently create a small web application with a few users.

> **The Right Large Language Model (LLM) for OpenHands**
>
> When trying out OpenHands, you potentially generate a large number of requests to a language model. If you use a cloud provider for this, it can quickly become expensive (we've been there too). OpenHands also supports local LLMs, but our attempts with llama3.1, codegemma, or deepseek-coder were very disappointing. None of them returned any useful results. Using proprietary models such as gpt-4o (the default setting in OpenHands), we were able to achieve small successes, which we want to present to you here.

11.1.1 Installation

To exploit the full potential of OpenHands, the program must also be able to install software. For example, if OpenHands is to create and test software in the Go language, it requires the compiler and the Go modules this program uses. If these components were to be installed on your computer's operating system at every attempt, your computer would soon be pretty messed up.

Docker containers provide an ideal solution in this context: OpenHands can do what it wants in a *sandbox container*. When you exit OpenHands, the container (with all the installed software) gets deleted. The working directory in which the desired code is created is retained, of course. We assume that you have basic experience with Docker and have installed it on your computer. You should also use a terminal window in which a standard Unix shell such as bash or zsh is running. On Linux and macOS, this shouldn't be a problem, whereas on Windows, you have to use Windows Subsystem for Linux (WSL) with a Linux system.

The current OpenHands version can be started as follows:

```
WORKSPACE_BASE=$(pwd)/workspace
docker run -it \
  --pull=always \
  -e SANDBOX_RUNTIME_CONTAINER_IMAGE=\
     ghcr.io/all-hands-ai/runtime:0.9-nikolaik \
  -e SANDBOX_USER_ID=$(id -u) \
  -e WORKSPACE_MOUNT_PATH=$WORKSPACE_BASE \
  -v $WORKSPACE_BASE:/opt/workspace_base \
  -v /var/run/docker.sock:/var/run/docker.sock \
  -p 3000:3000 \
  --add-host host.docker.internal:host-gateway \
  --name openhands-app-$(date +%Y%m%d%H%M%S) \
  ghcr.io/all-hands-ai/openhands:0.9
```

Note that a `workspace` folder will be created in the current directory in which your new software will be developed. If this folder already exists, it will be integrated together with the existing content. You can change this directory by adjusting the `WORKSPACE_ BASE` variable in the first line.

Including the `/var/run/docker.sock` socket allows the container to control the Docker daemon. OpenHands requires this setting so that the sandbox container can be started by the application. However, controlling the Docker daemon also gives the container access to all other Docker resources on your computer. So don't start OpenHands on a system on which important Docker applications are running productively.

At startup, a container is derived from the current Docker image, which is assigned the name `openhands-app-XXXXX`, where `XXXXX` is replaced with the current date and time. This kind of naming ensures that you can easily find the container of an aborted attempt at a later stage. The container contains all log files that were created during the test, which can be useful for analysis.

The web application then runs on *http://localhost:3000*. The sandbox container is started automatically as soon as you load the web interface and is assigned the name `openhands-sandbox-YYYYY`, where `YYYYY` stands for a randomly generated, unique ID. After the successful start, you can display the two containers with the Docker subcommand `ps` (the output has been specially formatted due to the long names):

```
> docker ps --format '{{.Image}} {{.Names}}'

  ghcr.io/all-hands-ai/runtime:0.9-nikolaik  openhands-sandbox...
  ghcr.io/all-hands-ai/openhands:0.9 openhands-app-20240910163059
```

11.1.2 The Web Interface

Once you've successfully completed the installation, you can start using OpenHands. To do this, open *http://localhost:3000* in your web browser with the OpenHands web interface.

The chat with OpenHands is created on the left-hand side. There, you enter your requirements, and OpenHands explains the steps that are carried out. You can view the generated code (and any other files and folders) in the top-right area of the browser window. Below this is an interactive terminal, which OpenHands itself also makes use of. This is a shell in a sandbox container.

For OpenHands to process your instructions, you must first configure the LLM and the *agent*, whereby the agent takes over the communication between the LLM and the rest of the software. The corresponding configuration dialog appears the first time the web interface is opened and can be called again at any time using the screw symbol at the bottom right.

Figure 11.2 The OpenHands Web Interface Currently Only Available in "Dark Mode"

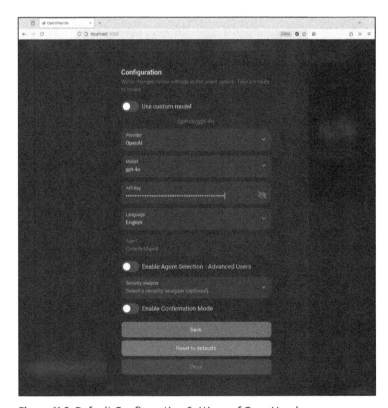

Figure 11.3 Default Configuration Settings of OpenHands

To use the default gpt-4o model from OpenAI, you must enter the API key that you've previously created in the settings of your OpenAI account (see Chapter 10). The default CodeActAgent is used as the agent for the gpt-4o model, which brings us to the topic of local LLMs. As already mentioned, our successes with local language models were extremely limited. The OpenHands online help also confirms this observation and refers to the *GPT-4* and *Claude 3* models as currently the best partners for OpenHands.

Perhaps this situation will have improved by the time you hold the book in your hands and you also want to give the more cost-effective variant with local models a chance. OpenHands supports the Ollama API, which we've already presented in Chapter 9, Section 9.6. For access to your local Ollama models to work, you must set the LLM_BASE_URL variable when starting the container:

```
docker run \
    ...
    -e LLM_BASE_URL="http://host.docker.internal:11434" \
    ...
    ghcr.io/all-hands-ai/openhands:0.9
```

If ollama isn't running on your local computer, but on a computer in the local area network (LAN; as was the case in our tests), you need to enter the Domain Name System (DNS) name of the computer on which the service is running instead of host.docker.internal. Make sure that the openhands app container has access to port 11434 on this computer and that no firewall is interfering. You can then enter the ollama/llama3.1 string in the configuration dialog under *Model* (if you want to use the llama3.1 LLM from your local installation). Note that switching to a model of a cloud provider, such as gpt-4o from OpenAI, only works if you restart the container and don't define the LLM_BASE_URL variable.

During our tests, we found the error messages in the browser chat not very helpful. OpenHands outputs a few technical error messages here, which often leads to very abbreviated, generic messages. The log output in the terminal window in which we started the Docker container was much more helpful. These messages are very detailed and usually provide quick information about the actual problem. Here is an excerpt from the log file of the Python web application:

```
CodeActAgent LEVEL 0 LOCAL STEP 13 GLOBAL STEP 13

06:37:31 - openhands:INFO: llm.py:486 - Cost: 0.04 USD |
  Accumulated Cost: 0.43 USD
Input tokens: 7757
Output tokens: 53

06:37:31 - ACTION
**CmdRunAction (source=EventSource.AGENT)**
```

```
THOUGHT: Let's use `netstat` to find the process using port 5000
  and then kill it.

First, let's find the process ID (PID) using port 5000.
COMMAND:
netstat -tuln | grep :5000
06:37:31 - openhands:INFO: runtime.py:359 - Awaiting session
06:37:31 - openhands:INFO: runtime.py:263 -
---------------------------Container logs:--------------------
    |INFO:     172.17.0.1:53856 - "GET /alive HTTP/1.1" 200 OK
    |INFO:     172.17.0.1:53856 - "POST /execute_action HTTP/1...
-------------------------------------------------------------
06:37:31 - openhands:INFO: session.py:139 - Server event
06:37:31 - OBSERVATION
**CmdOutputObservation (source=EventSource.AGENT, exit code=1)**
bash: netstat: command not found

[Python Interpreter: /openhands/poetry/openhands-ai-504_aCHf-p...
openhands@200f871dd6cf:/workspace $
06:37:31 - openhands:INFO: session.py:139 - Server event
```

For this reason, you should always keep an eye on the terminal window with these log messages when experimenting with OpenHands.

11.2   Using OpenHands

The first task for our artificial junior software developer is to create a web application with very simple user administration. The backend is to be programmed in Python, and the users are to be stored in an SQLite database. A website is used to manage users, where you can add new users and delete or change existing ones.

> **Prompt:** Create a JSON REST API backend using Python. Add a route health that responds with HTTP status code 200.

This task is of course child's play for a powerful model such as gpt-4o. OpenHands performs the following steps and only stumbles slightly:

1. The app.py file gets created. It contains the HTTP endpoint for /health and sets the web server port to 5000.

2. The python3 /workspace/app.py command starts the server as a background process.

3. The curl http://localhost:5000/health call fails with the error message, Failed to connect to localhost port 5000.

4. Due to the `lsof -i :5000` command, OpenHands attempts to check whether the server process is running on port 5000. This command also fails because `lsof` isn't installed in the sandbox container.

5. Because steps 3 and 4 were executed so quickly one after the other, there was no error message for the failed server start from step 2 in the log file. Only now does the error, `ModuleNotFoundError: No module named 'flask'` appear.

6. OpenHands fixes the problem with the installation of the Python module.

7. Calling `curl` again returns the expected result, and OpenHands returns control to the user.

After seven steps and a cost of $0.05, the first task is completed. A human programmer would probably not have made the mistake with the missing module installation in the first place, and, if they had, would have noticed the problem immediately after the server start.

OpenHands will now extend the application to include simple user management with a database in SQLite format.

> **Prompt**: Add code for a simple user administration, where users are stored in an SQLite database.

After several errors, all of which OpenHands can fix itself, there are two more routes to our backend. First, the list of all users can be called via GET. The second route is used to create new users by using POST. We've now reached 28 steps and a cost of $0.29.

Next, we want to have a web frontend for the application:

> **Prompt**: Add a web frontend where I can add a user to the database.

The input template on the very simple HTML page uses JavaScript without additional libraries, which is entirely to our liking. When submitting the form, the `fetch` function is used to send the entered data to the backend. OpenHands also extends the backend so that the website can be provided at the address, `http://localhost:5000/`.

However, we now also need a way to delete or edit the users.

> **Prompt**: Add code for deleting and editing users.

OpenHands only takes care of the backend and adds the routes with HTTP methods DELETE and PUT. The DELETE call is then tested with a nonexistent user ID, which results in an error message. OpenHands apparently considers this a successful test and proudly reports that the code has been added and tested. In the meantime, 48 steps have been completed and we've spent $0.59. The missing frontend code is added in just one step. As more tokens are required here, this step costs a little more.

> **Prompt**: Add frontend code so that I can see a list of all users with Edit and Delete buttons.

Finally, we want to create a few sample users so that we can try out the application.

> **Prompt**: Insert 5 example users.

OpenHands calls the REST API five times using `curl` and test user data, with each call costing $0.02. This is good to know if you want to add 10,000 test users, for example.

Figure 11.4 The "Finished" Web Application with a List of Users

Now you'll rightly say that you can also use ChatGPT or another chatbot in the browser for this simple example. After all, you also receive precise instructions on which files and folders need to be created.

11.2.1   Not Too Much at Once!

Actually, we would have liked to present you with a complete example with a prompt in this form:

> **Prompt**: Create a backend JSON REST API in Golang with a simple user administration. Users should have Email, Name, Password and get stored in an SQLite database. Add a frontend in vue.js where I can add, edit, and delete users.

The initial reaction from OpenHands is promising: a plan is presented in which the backend, including the database, is to be created first and then the frontend. But even the initialization of the Go module fails. If multiple attempts are made to resolve the problem, the folder for the Go module is created nested below the previous attempt. Unfortunately, the result was no better for the frontend code.

OpenHands stopped after 37 minutes and 60 steps without any significant result. Neither the desired backend in Go nor the frontend had any usable code. As this wasn't our first attempt that failed so miserably, we were at least able to use the time for a lunch break.

11.2.2   Conclusion

When we started working on the OpenHands sections and began to understand the technical concept behind it, our expectations were very high. The fascination with how well an LLM can deal with natural language and also with programming languages contributed to this. Disillusionment followed on its heels, however, and the failure of OpenHands due to trivial things, such as an incorrect path in the folder structure, leaves us with great doubts about the rapid implementation of an *autonomous software engineer*.

Nevertheless, OpenHands offers a very exciting outlook on how a part of software development could work in the future. This is still a very young project that will benefit massively from the further development of language models in particular.

We were most disappointed by the large discrepancy between local models and the OpenAI models. While local models for code completion or as a chat bot, achieve quite useful results, as we've presented in other chapters in this book, this was unfortunately not the case with OpenHands. If efficient software development is only possible in the future with access to one of the tech giants (including the inevitable data monitoring), then that would be a very sad development for us.

11.3   Aider

Aider inventor and developer Paul Gauthier describes his work somewhat succinctly as "Aider is AI pair programming in your terminal." In fact, Aider lacks a web or graphical user interface (GUI) interface. However, the rest of the description is an understatement from our point of view. Microsoft describes GitHub Copilot very similarly as "Your AI pair programmer."

With regard to functionality, however, Copilot and Aider are worlds apart! Aider expects relatively abstract instructions, automatically creates the required files or makes changes to existing project files, and finally asks whether you want to apply the changes as a Git commit. As we've already indicated in the introduction to this chapter, Aider acts much more independently than GitHub Copilot, Continue, or other plug-ins that are integrated into the editor or the development environment.

11.3.1 Installation

Aider is open-source software based on Python and Git—prerequisites that are almost certainly already fulfilled on your computer. To avoid package conflicts with other projects, it's best to install Aider in a virtual environment. We carried out our tests on macOS with Python 3.12.

```
mkdir aider    # Installation for macOS/Linux
cd aider
python3 -m venv .
source bin/activate
pip3 install aider-chat
```

On Windows, you need to use the python and pip commands instead of python3 and pip3. You also need to activate the environment using Scripts/Activate.ps1 instead of the source command. Updates can be carried out as follows:

```
pip3 install aider-chat --upgrade
```

For Aider to work, the program must have access to a language model. Assuming suitable hardware, a (strong!) local model is sufficient. We tested Aider with two commercial language models, GPT-4o from OpenAI and Claude Sonnet 3.5 from Anthropic. Sonnet 3.5 delivered noticeably better results in our tests.

By the time you read this book, there will be newer language models so we can't really tell which one will be best for Aider. But take a look at the following page, which summarizes which language model works how well with Aider: *https://aider.chat/docs/leaderboards*.

Now you need to define an environment variable for your API key, preferably in .bashrc or .zshrc on Linux or macOS. Alternatively, you can also save the key(s) in .env in the Aider environment directory.

To receive an API key, you must create an account with OpenAI or Anthropic and enter your contact and credit card details. You buy credits there in advance, which are then used up by API queries. About $10 should be sufficient for initial tests, so you don't have to invest a fortune to try out Aider.

```
# in ~./bashrc oder ~/.zshrc or aider/.env
export OPENAI_API_KEY='sk-xxxxx'
export ANTHROPIC_API_KEY='sk-yyyyy'
```

If one of the two variables is defined, Aider automatically recognizes the desired model. Once both variables have been defined, Aider opts for the Sonnet model from Anthropic. If required, you can explicitly select the language model with the `--4o` or `--sonnet` options.

11.3.2 Start

Aider is a terminal tool, but that doesn't mean you have to make do without your editor. A good way to use it is to open a project directory in your favorite editor (we used Visual Studio Code [VS Code]). For a new project, this is initially empty. Then, open a terminal window in the editor, and activate the Aider virtual environment. Make the project directory the active directory.

```
cd ~/aider           # change to Aider directory
source bin/activate  # activate virtual environment
cd ~/myproject       # change to the project directory
```

Now start the AI tool using `aider`. If there is no Git repository in the local directory, Aider recommends creating one. All you have to do is confirm the corresponding query.

```
aider
```

```
  Aider v0.50.1
  Models: claude-3-5-sonnet-20240620 with diff edit format,
   weak model claude-3-haiku-20240307
  Use /help <question> for help, run "aider --help"
    to see cmd line args
```

11.3.3 Aider Commands

For new projects, you can start directly by giving the tool instructions, for example, which task your new program should perform or which programming language you want to use. Aider creates the required new files itself. To run and test the program, you can either use the editor, a development environment, or simply a second terminal window.

Of course, you can also use Aider to edit an existing project with many files. Aider commands such as `/command` play an important role in this type of work (see Table 11.2). You can find a complete reference at *https://aider.chat/docs/usage/commands.html*.

While the created files are added directly to the context for new projects, you must explicitly add the files to be edited to the context for existing projects via /add or /read or remove them again by using /drop. The difference between /add and /read is that Aider only makes changes in the /add files during further prompts, while it considers the /read file only as a reference. Files passed with /read can, for example, contain the code of a class whose methods you want to use, documentation, or schema files.

A major problem with all language models is that their information status ends with the date on which the training material was created. This is a huge limitation, especially if you use modern programming languages or libraries.

Aider gets around this problem with a simple concept: you can use /web to transfer the address of a website. Then, Aider downloads this page, converts the text found there into Markdown format, and inserts it into the context.

Command	Meaning
/add file	Adding a code file to the context
/ask	Asking questions about the code (no changes)
/clear	Deleting the chat history (files remain in context)
/diff	Showing changes from the last commit
/drop file	Removing a code file from the context
/exit or /quit	Exiting Aider
/git cmd	Executing a Git command (e.g., /git status)
/help myquestion	Asking the language model a general question
/help /cmd	Displaying help for the specified command
/lint	Checking the syntax of current files, fixing errors
/ls	Showing files in the context
/map	Showing a repository overview
/read	Adding a file to a context (read-only)
/run cmd	Executing a shell command (e.g., /run cd subdir)
/undo	Revoking the last commit
/web url	Reading a file and adding it to the context

Table 11.2 Important Aider Commands

Less Is More

It's not a good idea to add as many or even all files of a large project to the context! This is also true for the number of pages transferred from an online documentation using /web.

We've already mentioned it several times in this book: although modern language models have pleasingly large context windows, they struggle to separate the essential from the nonessential. Too much contextual information often unbalances the language model. The choice of what is important or not so important is up to you!

11.3.4 Multiline Prompts

As soon as you press [Enter], Aider considers the entry to be complete and starts responding. Multiline prompts can be entered via [Alt]+[Enter]. However, experience shows that there is a high risk of forgetting [Alt]. The second approach therefore seems more helpful to us: start the entry with a curly bracket, and then press [Enter]. You can then enter as many additional lines as you like and press [Enter] before ending your prompt with another curly bracket.

```
> {
> this is a long question
> about the meaning of life
> }
```

11.3.5 Coding Rules and Conventions

It often happens that predefined conventions for the code must be adhered to in a project. In this case, the Aider documentation recommends saving the rules as ordinary text instructions in a Markdown file in the project directory. You can add this file to the context with /read or pass it as a parameter when starting aider:

```
aider code-conventions.md
```

11.3.6 Aider in the Web Browser

Aider sees itself as a terminal tool. The program can also be operated in the web browser. To do this, you need to start the program with the additional --browser option. A new tab with the Aider web interface will then appear in your web browser.

The Aider manual describes this function as experimental, and we can confirm this impression: the interface is visually appealing, but despite asking questions in English, Aider formulated the answers in German (the default language set in our web browser), made changes to the code despite /ask, didn't understand commands such as /undo, and so on.

11.4   Using Aider

After this introduction to the basic functions, we took a look at how Aider performs in real-life work. Our conclusion is a little like the basic tenor of this book: it depends on the complexity of the project, the task, the programming languages and libraries used, and so on.

11.4.1   Example 1: Sorting Photos

The following example shows you what using Aider can look like for a new project. You want to develop a Python script that moves photos to directories of type 2024/09 based on the date they were taken.

Prompt: I have a directory with photos (JPEGs with EXIF). I need a Python script to move the files to a new location according to their creation date, with year and month as subdirectories, for example, destination-directory/2024/08/myphoto.jpg. The source and destination directories are passed as arguments. If the destination directory is empty, create it.

Certainly! I'll create a Python script that accomplishes this task. Because you haven't shared any existing files, I'll create a new file called organize_photos.py with the required functionality. Here's the script.

[50 lines of code, more explanation]

To use this script, you'll need to install the Pillow library:

```
pip install Pillow
```

The script looks good at first glance. To easily set up a Python environment for testing, ask for a requirements file.

Prompt: Please create requirements.txt.

Now, open a second terminal window or a tab, and set up the test environment there. (You can, of course, also create the environment in a directory other than the project directory if you prefer.)

```
python3 -m venv .
source bin/activate
pip3 install -r requirements.txt
python3 organize_photos.py ~/unsorted-pictures ~/pictures-by-date
```

> **Tip: Work in Small Steps**
>
> Even if Aider acts much more independently than GitHub Copilot, for example, there is a common denominator between current AI tools, regardless of their level. Prompts with too many or too complex tasks overwhelm Aider. Develop your project or solve your task in small steps. Perhaps it gives you some satisfaction that AI tools, for all their actual or ostensible "intelligence," are still frighteningly limited in this respect.

The script works straight away. Great! Now you want to add some features to the script. First of all, you want an option so that the script doesn't move the photos but copies them. This is very useful, especially for testing purposes.

Prompt: Add an option to copy instead of move the photos.

Certainly! I'll modify the `organize_photos.py` script ...

[code changes and explanation]

These changes will allow the user to specify whether they want to copy or move the photos. The script can now be run like this:

```
python organize_photos.py <source_directory> \
        <destination_directory> copy
```

or

```
python organize_photos.py <source_directory> \
        <destination_directory> move
```

Aider has implemented the option as an additional parameter that you should specify after the source and destination directories. That doesn't make you happy because you were thinking more of a parameter like `--copy` as in Linux or macOS. Aider isn't clairvoyant, but the tool isn't offended if it has to rebuild the code in a further step.

Prompt: Please use a Linux-style option (`--copy`) that can be passed at any position.

Aider automatically performs a syntax check after each change to detect errors. In this case, Aider added `sys.exit` to the code, but obviously forgot `import sys`. Aider asks whether it should try to rectify the error.

```
Running: /Users/kofler/no-sync/aider/bin/python3.12 \
  -m flake8 --select=E9,F821,F823,F831,F406,\
  F407,F701,F702,F704,F706 --show-source \
  --isolated organize_photos.py
```

```
organize_photos.py:55:9: F821 undefined name 'sys'
sys.exit(1)

Attempt to fix lint errors? y
```

The error is trivial, and Aider manages to correct it. Then, you return to your test terminal and try out the new --copy option. Everything works fine. Now you ask for further improvement such as two additional options and the use of a separate directory for photos without EXIF data. (So far, the script has used *File Modification Time* to sort such photos, but this is rarely useful.)

> **Prompt:** Add two more options: --quiet to suppress output (but show error messages) and --help to show a help text; also add short options -c for --copy, -h for --help, and -q for --quiet.

> **Prompt:** Change the handling for files without EXIF data or with invalid EXIF data: move/copy these files into a special directory no-exif-data; also show a warning, even when --quiet is set.

You take some of the photos with your iPhone that uses the High Efficiency Image Coding (HEIC) format as standard. Wouldn't it be great if your script could cope with this type of image?

> **Prompt:** Next, add support for HEIC files.

> To add support for HEIC files, we need to make a few changes to the organize_photos.py file. We'll need to add the pillow-heif library to handle HEIC files, update the file extension check, and modify the get_creation_date function to work with HEIC files. Here are the necessary changes ...

To test this in the second terminal, you must first repeat pip3 install -r requirements.txt to load the additionally required pillow-heif module into the virtual environment. After that, problems occur for the first time: the script indicates that it can't find any EXIF data in the HEIC files. You use an image editor to check this. The EXIF tags are definitely present. But pillow-heif apparently has problems reading them.

Now you've reached the moment when you have to deal with the code for the first time. (Keep in mind that you have *not* touched a *single line* of code yet! Up to this point in the example, you wouldn't have needed an editor at all.)

Your first idea is to check whether Aider is using current versions of pillow and pillow-heif. This isn't the case. You research on the internet, enter the latest version numbers in `requirements.txt`, repeat `pip3 install`, and test the program again. Unfortunately, there is no improvement.

Outdated Training Material

Old version numbers are a typical problem with common language models. The language model in this test from Anthropic was trained with old code and therefore often uses old versions. You should always take a look at version numbers in requirements files, Dockerfiles, or other text files that specify software versions! Update these numbers to the latest version to avoid possible security problems as well.

At this point, a new phase of program development begins: troubleshooting. You point out the problem to Aider and ask for it to be rectified.

> **Prompt:** The script doesn't work for HEIC files (shows warning "no valid EXIF data ..."); please try to fix it.

Aider builds a case distinction into the code that differentiates between JPEG and HEIF files. However, both code branches contain exactly the same code, so the faulty behavior doesn't change. Use /undo to revoke the last commit. The response to your succinct statement that the fix was ineffective is surprising. Aider inserts unsolicited logging code to give you a chance to get to the bottom of the error.

> **Prompt:** /undo

> **Prompt:** Your last fix didn't work.

> I apologize for the confusion. Let's try a different approach to fix the issue with HEIC files. We'll modify the `get_creation_date` function to handle both JPEG and HEIC files more explicitly. Here's the proposed change ...
>
> Additionally, let's add some debug logging to help identify any issues ...

You test the program again and take a look at the numerous issues. It turns out that `pillow-heif` only sees a fraction of the approximately 30 EXIF attributes. The DateTime-Original attribute (tag 36867) parsed by the script is missing, but at least DateTime (tag 306) is visible. This attribute specifies the date and time at which the image was last changed. Many pictures are never changed so that the information matches the date on which they were taken.

Thus, you ask Aider to parse DateTimeOriginal first, but to consider DateTime in a second step if necessary. (At this point, we want to state that the error in reading the EXIF data was caused by shortcomings in the Pillow library or incompatibilities in the HEIC format, not by Aider. If you had developed the code without AI tools, you would have stumbled at this point too.)

> **Prompt:** Apparently, the script can only access a limited set of EXIF properties for HEIC files; first try to read 36867; if this fails, use tag 306 instead.

> **Prompt:** This works; now remove the debugging code.

Up to this point, Anthropic has incurred costs of $0.26, which is manageable. Including debugging, the time required was just under 90 minutes. The final script consists of 85 lines of code.

In our opinion, the remarkable thing about this example isn't so much the result (if you're interested in the code, take a look at the sample files in this book), but rather the way to get there. We could have written the script ourselves in an editor with ChatGPT or Copilot support. But instead, we really delegated all the coding and just took care of the testing. This is an entirely new approach. Of course, you don't have to work exclusively in this way: you still have full control over your code and can change or extend it at any time before asking Aider for help with a new task.

Aider offers another advantage compared to conventional program development: the entire process is wonderfully documented, first by a log of all prompts and outputs in the .aider.chat.history.md file in the project directory, and second, by an easy-to-read Git message sequence:

```
git log --pretty=format:"%s" --reverse
```

```
feat: Implement photo organization script
feat: add requirements.txt
feat: add option to copy or move photos
feat: Add Linux-style --copy option to organize_photos script
fix: import sys module
feat: Add --quiet and --help options to organize_photos.py
feat: Handle files without EXIF data or with invalid EXIF data
feat: Add support for HEIC files
fix: Move `register_heif_opener()` call after its import
fix: Handle HEIC files in get_creation_date function
feat: Add fallback to DateTime EXIF tag for HEIC files
refactor: remove debugging code
```

11.4.2   Example 2: Web Scraping

Before you think that Aider is the new miracle tool, we want to bring you back down to earth with the next example. Our goal is to program a script that extracts the top five bestsellers from *www.sap-press.com* by means of web scraping, which actually isn't a particularly difficult task. First, we ask Aider to use /web to load the bestseller page and display the five titles. The only purpose of this is to check whether Aider can handle the website at all.

> **Prompt:** /web *https://www.sap-press.com/bestsellers*

> **Prompt:** List the top five titles of *https://www.sap-press.com/bestsellers*.

1. Materials Management with SAP S/4HANA: Business Processes and Configuration
2. Cloud Integration with SAP Integration Suite: The Comprehensive Guide
3. Developing AI Applications: An Introduction
4. SAP Build: No-Code Development, Centralized Access, and Process Automation
5. SAP Signavio: Business Process Transformation

As we can see, Aider does understand the structure of the website. Now we want Aider to automate this task using a Python script.

> **Prompt:** Please write a Python script to extract the top five bestseller titles from *https://www.sap-press.com/bestsellers*.

> **Prompt:** Add a requirement file to the project.

In VS Code, we take a look at `bestseller_scraper.py` and `requirements.txt`. Both files look good. We set up a virtual environment again, install the modules, and test the program. Unfortunately, the result is disappointing: the script only outputs a heading, while the actual list remains empty. So, we ask for improvements.

> **Prompt:** `bestseller_scraper.py` only outputs an empty list; please try to fix the code.

> I understand that the `bestseller_scraper.py` is outputting an empty list, and we need to fix the code. Let's analyze and modify the script to address this issue. The problem likely lies in how we're selecting elements from the web page. We'll adjust the HTML parsing to correctly identify and extract the book titles.
>
> [new code]

Aider rebuilds the code a little and changes the names of the HTML classes the script searches for. To cut a long story short: Aider fails again, as the script still returns an empty list. This is disappointing insofar as Aider can read, parse, and understand HTML pages via the /web command. However, this functionality relates to the content of the page rather than its HTML structure. After a few more attempts, we abort this test without any success. Aider provides no advantage here compared to "honest," in-house program development.

API Errors

The HTML comprehension problems described here are unfortunately not the only annoyance we encountered during our Aider tests. It happened relatively often that our prompts were answered with error messages indicating that the AI backend we were using (Anthropic) was currently overloaded or didn't respond for some other reason. It usually helped to simply run the prompt again.

Of course, such API errors are most likely not the fault of Aider, but have to do with shortcomings of the respective AI provider. However, we've never had as many issues with any other tool presented in this book as with Aider. That's odd!

11.4.3   Example 3: Implementing Changes in a Large Project

Most recently, we tried to use Aider to make changes to a complex, real-world project consisting of several hundred files (PHP, JavaScript, Cascading Style Sheets [CSS]). The following paragraphs show a few exemplary prompts. Our goal was to change details of an HTML table that gets output by functions of the `statistics.php` file. (All functions mentioned in the prompts can be found in this file.) As the data displayed in the HTML table originates from a database, we've also adopted the database schema in the context.

Prompt: `/add accounts/statistics.php`

Prompt: `/read doc/accounting-schema.sql`

Prompt: In `getUsageStatistics()`, add exception handling for the call `connectToCustomerDatabase()`; if the connection fails, continue loop with the next customer.

Prompt: `showUsage()` displays the customer name in the first column of an HTML table; make this name a link to `editCustomers.php?id=nnn` where `nnn` is the customer ID.

Prompt: Still in the first column of the HTML table created by `showUsage()`: after the link, add the year of the initial order; you get the necessary data from the database column `customers.initialOrderDate`.

Our findings were mixed, not totally bad, but not great either:

- We got the impression that only prompts that are very carefully and precisely formulated lead to the goal. Sometimes we wondered what takes longer: the Aider prompt input or a direct modification of the file with the support of a wizard integrated into the editor.

- Even with precisely formulated prompts, Aider didn't always do exactly what we expected. Sometimes, we revoked the entire commit using /undo and then tried again with a better prompt. At other times, we simply carried out the corrections ourselves, which often consisted of simple mix-ups of column or variable names. Of course, it's remarkable when Aider understands an instruction 90% of the time and carries it out correctly. But that still doesn't suffice for really efficient work.

Long story short is that purely instruction-oriented programming takes some getting used to, but is certainly exciting. However, we didn't have the feeling that Aider would save us much time. The benefits of this approach are probably greatest when you develop new code from scratch—be it a completely new project or a new method or class in an existing project.

11.4.4   More Prompting Examples and Video Tips

By default, Aider saves the prompt history, including all responses in the `.aider.chat.history.md` Markdown file. A collection of mostly short examples derived from this (each prompt plus response in visually appealing formatting) can be found at *https://aider.chat/examples/README.html*.

There is also a whole series of videos available on the internet showing Aider at work. Aider developers have compiled links to such videos at *https://aider.chat/docs/usage/tutorials.html*.

These examples show various aspects of the use of Aider and provide many application ideas. However, we had the impression that some examples were really optimized for Aider and therefore conveyed an overly optimistic picture. As with all other AI tools presented in this book, you must also expect problems and errors when using Aider in real life.

Chapter 12

Retrieval-Augmented Generation and Text-to-SQL

This chapter differs somewhat from the rest of the book, as we're not presenting working techniques and tools that make programming easier for you, but rather developing a program that uses AI to achieve a result. Of course, there is still a connection: the code for the examples in this chapter was again created with AI support. So, this chapter again contains a lot of prompts as well as some comments on why we formulated the prompt in exactly the same way and why some of the previous attempts failed.

The problem for the following AI application is to analyze data from different sources using a large language model (LLM). We use PDF documents, web pages in HTML format, and an SQL database as basic data. In addition, two different techniques are used to realize these requirements: retrieval-augmented generation (RAG) and Text-to-SQL. The idea is to use these technologies to make a dormant treasure trove of data searchable.

In numerous projects, the keyword search for archived PDFs and other documents isn't sufficient, let alone the data in an SQL database. Asking questions in natural language and having them answered with the content of this data can really be a game changer in certain areas.

Retrieval-augmented generation (RAG) is one of the buzz words in the current AI hype. Because it's impossible to add your own information to a ready-made LLM with manageable effort, an additional small, custom language model, the *embedding* model, is placed in front of a large, general language model. When a query is made, the results of this model are redirected to the LLM, which then generates the response for the user.

The embedding model converts unstructured data, such as text or images, into multi-dimensional vectors. As with a geographical coordinate system that describes points on the earth, these vectors provide information about the position of information in relation to each other. However, these vectors aren't limited to three dimensions as in a geographical coordinate system, but rather can have hundreds or thousands of dimensions.

Using Text-to-SQL, an LLM generates an SQL query from a query in natural language. For this to work, information on the database structure is transferred to the LLM along with the request. As you can imagine, the LLM only has a chance of formulating SQL queries that make sense if the table and column names have meaningful designations.

For the development of the AI application in the following example, we use the open-source library, *LlamaIndex* (*www.llamaindex.ai*). This library provides connections for the Python programming language as well as for TypeScript. We've chosen Python for this example.

12.1 RAG Quick Start

As an introduction to the topic, we want to show you how quickly you can get an executable AI application if the AI writes it itself. We want to ask questions on an HTML page about documents we store in a `pdfs` folder on the hard disk. These are annual reports and further information on a research project on insect censuses in Austria, which we'll discuss in more detail later in the chapter. Our prompt to the Claude 3.5 Sonnet LLM reads as follows:

> **Prompt:** Generate a FastAPI backend for a LlamaIndex Q&A application and an HTML page to input questions. Data for LlamaIndex is in a folder `pdfs`. Don't use templating, but serve the HTML file from server root.

Unfortunately, the generated Python code isn't executable because the LlamaIndex library has changed somewhat since the end of the training data for Claude. This is an issue we've already encountered several times in this book. However, the changes are minimal (the `GPTSimpleVectorIndex` has been renamed `VectorStoreIndex`), and after changing three lines of code, the application runs. Here, we show you the entire backend script that starts the web server, answers questions using AI, and returns the answer in JavaScript Object Notation (JSON) format:

```python
from fastapi import FastAPI, HTTPException
from fastapi.staticfiles import StaticFiles
from pydantic import BaseModel
from llama_index.core import (
  VectorStoreIndex, SimpleDirectoryReader
)

app = FastAPI()

documents = SimpleDirectoryReader("pdfs").load_data()
index = VectorStoreIndex.from_documents(documents)
query_engine = index.as_query_engine()

class Question(BaseModel):
    text: str
```

```
@app.post("/api/ask")
async def ask_question(question: Question):
    try:
        response = query_engine.query(question.text)
        return {"answer": str(response)}
    except Exception as e:
        raise HTTPException(status_code=500, detail=str(e))

app.mount("/", StaticFiles(directory="static", html=True),
        name="static")

if __name__ == "__main__":
    import uvicorn
    uvicorn.run(app, host="0.0.0.0", port=8000)
```

Most of the few lines of code relate to the web server, which can easily be set up using the FastAPI and Uvicorn libraries. HTML documents in the static folder are provided regularly via the web server. In the case of an HTTP POST request sent to the /api/ask URL, the content of the request gets mapped to the previously defined Question class. The text entry it contains is passed to the query_engine call.

This is where the magic happens: with just three lines of code, the LlamaIndex library creates the query_engine, which can search the content of PDF documents with queries in natural language. To do this, the SimpleDirectoryReader reads a folder with files, in our case, PDF documents. A VectorStoreIndex is then generated from these documents, from which the query_engine is derived.

We'll explain what happens in detail during these steps in the following sections. First, we want to try out our application. As Claude explains in the short explanation, we need to install the necessary Python libraries:

```
pip install fastapi uvicorn llama-index pydantic
```

The HTML frontend requires just a few lines of JavaScript code and no additional libraries. We don't need to make any changes to the file that we save as index.html in the static folder. The central part of the HTML file reads as follows:

```
<textarea id="question" placeholder="Enter your question here">
</textarea>
<button onclick="askQuestion()">Ask</button>
<div id="answer"></div>

<script>
```

```
async function askQuestion() {
  const question = document.getElementById("question").value;
  const answerDiv = document.getElementById("answer");
  answerDiv.innerHTML = "Loading...";
  try {
    const response = await fetch("/api/ask", {
      method: "POST",
      headers: {
        "Content-Type": "application/json",
      },
      body: JSON.stringify({ text: question }),
    });
    if (!response.ok) {
      throw new Error("Network response was not ok");
    }
    const data = await response.json();
    answerDiv.innerHTML = data.answer;
  } catch (error) {
    answerDiv.innerHTML = "Error: " + error.message;
  }
}
```

```
</script>
```

The askQuestion function, which gets executed when the **Ask** button is clicked, sends the content of the text field with the question ID as a JSON string to the backend using an HTTP POST request. If the request is successfully answered, the content of the data.answer JSON structure gets inserted in the corresponding HTML-DIV area.

What Claude didn't mention in the description is that LlamaIndex swaps out the tough work to OpenAI. Thus, the application can only work if you have an OpenAI API key and available credit in your account. To give the application access to your account, you must set the API key in the OPENAI_API_KEY environment variable. In common Linux shells, you need to call the following command:

```
export OPENAI_API_KEY="sk-proj-xxxxxxxx"
```

Now we can start our application using the following command:

```
python main.py
```

The result is quite convincing. After 10 minutes, the first web application is ready, answering questions in natural language about PDF documents on the solid-state drive (SSD).

Before you close the book (or switch off the e-reader) and get to work implementing this example, we want to point out a few shortcomings of this application:

- Each time you start the web server, the index gets regenerated. This, however, isn't necessary because you can also save a `VectorStoreIndex` locally.

- The default settings require you to have access to OpenAI and pay for it. However, LlamaIndex works with other cloud providers and also with local LLMs, and it's not at all bad at that.

- You can set very precisely which types of documents you want to index and how the indexing works in detail. There is great potential for optimization here.

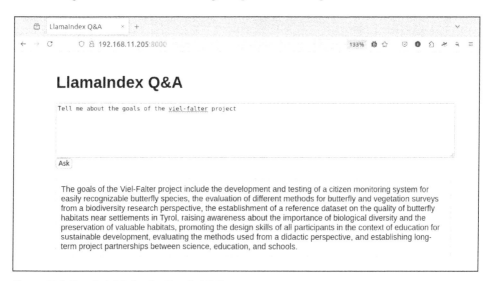

Figure 12.1 Our Quick Introduction to RAG

When reading the following sections, you'll get a deeper insight into how RAG and the LlamaIndex library work.

12.2   The Viel-Falter (Many Butterflies) Project

Instead of a Hello World example, we want to use data from a real project for this example. We'll see whether we can achieve added value with our AI application.

The viel-falter (translated: *many butterflies*) project is an Austrian initiative to monitor butterflies as indicators of changes in the natural and cultural landscape. The monitoring is managed by the University of Innsbruck. One of the authors has been in charge of the database, the input options, and the internal administration tools for the project for more than 10 years.

We've chosen this project for our example because, first, we have the option of making documents searchable that can't be found in one of the large LLMs (due to RAG), and, second, we can also demonstrate access to an SQL database via natural language (Section 12.7).

Figure 12.2 Part of the Website of the Viel Falter Project (https://viel-falter.at)

As the scope of the project is limited due to its regional nature, we weren't very surprised that the following prompt didn't lead to any usable results with the actually very good Claude 3.5 Sonnet LLM:

Prompt: Tell me about the goals and the partners of the viel-falter project.

I apologize, but I don't have specific information about a project called "viel-falter" in my knowledge base.

The latest information can be found on the internet at *https://viel-falter.at*, but we didn't expect this website to be included in Claude's index. To enable our AI wizard to provide competent answers, we'll use an embedding model to convert website content and some PDF documents into vectors. The results of this process will be saved in a vector database that allows for a search at a later stage.

12.3 Loading Documents

The annual reports in PDF format contain valuable information about the project and can be found on the WordPress website at *https://viel-falter.at/ergebnisse/veroeffentli-chungen*. To avoid having to download each of the 18 PDF documents individually, we've once again made use of AI:

> **Prompt:** How do I save linked PDF documents from the website *https://viel-falter.at/ergebnisse/veroeffentlichungen* to a local folder pdfs using Linux shell commands? Don't use temporary files but Unix pipes and remove duplicates. This site uses only absolute URLs.

We've tweaked the prompt a little because *Claude 3.5 Sonnet* initially created temporary files, which we don't consider necessary. Apart from the superfluous -p when creating the directory in the first line (the switch would only be necessary if you were creating a nested directory structure), there isn't much to criticize:

```
mkdir -p pdfs && \
wget -qO- https://viel-falter.at/ergebnisse/veroeffentlichungen/\
  | grep -oP 'href="[^"]*\.pdf"' | \
    sed 's/href="//;s/"$//' | \
    sort -u | \
    xargs -I {} wget -qnc -P pdfs {}
```

The wget command in line 2 loads the HTML code of the web page, which is forwarded to the grep command by the pipe character. The -o switch is used here to output only the part of the line in which the regular expression was found. The setting for Perl-compatible regular expressions (-P) wouldn't be necessary in this case but is also not an error. sed is used to remove the preceding href= string and the quotation marks from the links. sort -u removes the duplicates, and finally xargs executes a wget command for each link found, with -P specifying the path for saving the documents.

Without indicating that all URLs are absolute, Claude generates an address with a preceding protocol and host name (*https://viel-falter.at/*) in the last step, but points out in the explanatory text that this procedure will only work for relative links.

The command used here doesn't necessarily have to work just as well on another website. The search term in the grep command uses double quotation marks, which isn't actually required in HTML; however, the WordPress content management system uses double quotation marks.

Other than the PDF documents, we'll also include a list of websites in our embedding model. These are downloaded from the internet and converted into text when the application gets started.

12.3.1   SimpleDirectoryReader and SimpleWebPageReader

The LlamaIndex library uses *documents* and *nodes* as essential components. Documents are containers for data, such as a PDF document or a Markdown document. However, the output of a database query or the response from a web server can also be a document. Metadata is stored for a document (e.g., the file name of a file) and relations to other documents (e.g., the folder on the hard disk). Nodes are parts of a document that also contain metadata and relations.

LlamaIndex makes it very easy for us to convert the downloaded PDF documents into the internal document format:

```
from llama_index.core import SimpleDirectoryReader

pdf_documents = SimpleDirectoryReader("pdfs").load_data(
  show_progress=True
)
```

The `SimpleDirectoryReader` from the `llama_index.core` package generates a list of all files in the `pdfs` folder in the LlamaIndex document format. In our case, these are 18 PDF documents. The `show_progress=True` switch creates progress bars in the command line. As an extension to the example in Section 12.1, we'll also use some important websites as documents. For this purpose, we need the `SimpleWebPageReader` library:

```
from llama_index.readers.web import SimpleWebPageReader

webpages = [
  "https://viel-falter.at/",
  "https://viel-falter.at/monitoring/ziele-vision/",
  "https://viel-falter.at/projektpartner/",
  "https://viel-falter.at/news/",
  "https://viel-falter.at/tagfalter-exkursion/",
];

web_documents = SimpleWebPageReader(
  html_to_text = True).load_data(webpages);
```

The `html_to_text` option specifies that we're only interested in the textual content of the web pages. Both `pdf_documents` and `web_documents` now contain a list of LlamaIndex documents, which we'll convert into a searchable index in a further step. To combine the two lists, we use the + operator in Python:

```
documents = pdf_documents + web_documents
```

However, the `SimpleDirectoryReader` and the `SimpleWebPageReader` are just two examples of ways in which you can load data into LlamaIndex. On the LLamaHub (*https://llamahub.ai/*) website, you'll find Python modules for countless other formats.

12.4   Creating an Index

As we saw in Section 12.1, LlamaIndex provides a very convenient function for converting documents into a searchable index: `VectorStoreIndex.from_documents`. The function merely requires the list of documents as a parameter. LlamaIndex performs some interesting steps in the background, which we'll take a closer look at in this section.

First, LlamaIndex nodes are created from the LlamaIndex documents. For this purpose, a text splitter is used such as the `SentenceSplitter`. In contrast to an "ordinary" text splitter, this one attaches importance to not splitting a sentence into multiple nodes.

To give you a feeling for how the splitting of documents into nodes using the `SentenceSplitter` can look in numbers, we've counted the PDF pages from the 18 documents in our example (with the help of a Linux command that we had created by the Claude AI tool, of course) and compared them with the generated nodes. The 378 PDF pages become 466 LlamaIndex nodes. We've printed parts of the content of such a node without embedding in the form of a JSON structure for you here:

```
{
  "page_label": "5",
  "file_name": "Vielfalter-Falter_Tirol_2018_bis_2022.pdf",
  "file_path": "/data/pdfs/Vielfalter-Falter_Tirol_2018_bis_...",
  "file_type": "application/pdf",
  "file_size": 3190932,
  "creation_date": "2024-07-23",
  "last_modified_date": "2024-07-23",
  "_node_content": "{\"id_\": \"2d677c1e-da77-459d-9282...\",
     \"text\": \"Im Expert*innen-Monitoring werden inner ...",
     \"class_name\": \"TextNode\"}",
  "_node_type": "TextNode",
  "document_id": "94bdaa9f-c86c-4cf2-b38b-6d03805d265c",
  "doc_id": "94bdaa9f-c86c-4cf2-b38b-6d03805d265c",
  "ref_doc_id": "94bdaa9f-c86c-4cf2-b38b-6d03805d265c"
}
```

You'll recognize some metadata about the document. The content (text) is assigned to the `_node_content` key.

You can control this transformation of documents into nodes via the additional `transformations` parameter when calling `VectorStoreIndex.from_documents`. Then, a list of functions to be processed gets transferred:

```
from llama_index.core import VectorStoreIndex
from llama_index.core.extractors import TitleExtractor
from llama_index.core.node_parser import SentenceSplitter
from llama_index.embeddings.openai import OpenAIEmbedding
```

```
index = VectorStoreIndex.from_documents(
    documents=documents,
    transformations=[
        SentenceSplitter(),
        TitleExtractor(),
        OpenAIEmbedding(),
    ],
    show_progress=True,
)
```

After the `SentenceSplitter`, we use the `TitleExtractor`, which is another module from the LlamaIndex library. It tries to generate a short summary of the split text with the help of an LLM. (We'll show you how to set the LLM a little later.) The process significantly slows down indexing because the LLM is queried for each of the 466 nodes but also leads to better results in the subsequent search.

Note that LLMs in the cloud usually have limits on how many API calls they can make in a certain period of time. At Anthropic, the API limits for the Claude 3.5 Sonnet model we used were 50 requests and 40,000 tokens per minute. The limit of requests alone is quickly exceeded with the 466 requests that are executed within a few minutes. We therefore used the `TitleExtractor` together with a local Ollama instance, where the limits are only specified by its hardware.

The third entry in the list of `transformations` refers to the embedding model. This is where it gets really exciting because the model has a major influence on the generated index, and the choice of models is almost endless.

12.4.1   The Embedding Model

We use embedding models to convert text passages from the PDF documents and the loaded web pages into vectors. However, a vector doesn't represent a word or a token, but an entire paragraph or a whole page. We'll show you an example of what such a vector can look like. Depending on how large the training data was and how many dimensions the vectors have, embedding models can range in size from a few hundred megabytes to several gigabytes. The derived vectors that we'll save later, on the other hand, only take up a few megabytes for our dataset.

The `OpenAIEmbedding()` call shown previously loads the proprietary `text-embedding-ada-002-v2` model from OpenAI. To use it, you need an API key and credit in your OpenAI account. However, you can also use an open-source model from the Hugging Face platform:

```
from llama_index.embeddings.huggingface import (
    HuggingFaceEmbedding
)
```

```
embed_model_hf = HuggingFaceEmbedding(
    cache_folder="/home/user/huggingface_cache",
    model_name="BAAI/bge-m3",
)
```

It makes sense to specify the `cache_folder` for Hugging Face, as otherwise the model has to be reloaded from the internet again and again. Even if embedding models aren't as large as local LLMs, they quickly add up to a few gigabytes.

The `BAAI/bge-m3` used here has been trained with more than 100 languages and is available under the free MIT license. The vectors have 1,024 dimensions, and the model can process inputs with up to 8,192 tokens. However, the memory requirement of over 8 GB shouldn't be underestimated. If you only work with English texts, the much smaller `BAAI/bge-base-en-v1.5` model is a good alternative.

Hugging Face, the Platform for AI Models

At *https://huggingface.co* you'll find a large number of models as well as the *Massive Text Embedding Benchmark* leaderboard (*https://huggingface.co/spaces/mteb/leaderboard*), a ranking of current models with their parameters.

If you're looking for the right model for your purpose, you can filter by topic or keyword here. You'll also find training data for many of the free models.

If you operate a local Ollama server, you can also install embedding models there and use them for index generation:

```
from llama_index.embeddings.ollama import OllamaEmbedding
```

```
embed_model_ol = OllamaEmbedding(model_name="mxbai-embed-large")
```

Here, `mxbai-embed-large` uses 1,024 dimensions and has a memory requirement of 670 MB. In our tests with PDF documents and websites, the model provided useful results, although not as good hits as the much larger `BAAI/bge-m3` from Hugging Face.

But what exactly is the embedding model used for? The input data is compared with the trained model data, and the vectors for the input data are derived from this. The embedding, that is, the vector, for the LlamaIndex node shown previously looks as follows, for example:

```
[ 0.010941985063254, -0.007302129175513, 0.017867475748062, ... ]
```

We've refrained from printing the remaining 765 floats. The open source `BAAI/bge-base-en-v1.5` model, which we used in this experiment, uses 768 dimensions.

Creating the index is a computationally intensive operation. Especially if you use local models (in Ollama or from Hugging Face), a powerful graphics processing unit (GPU) will make a noticeable difference. Modern graphics cards from NVIDIA or the processors in modern Apple computers currently provide the best support. The compute time also depends on the model used. Multiple dimensions and larger models require more computing time but also provide better results.

For embedding models from the cloud, you don't need any powerful hardware yourself because that work takes place in the data centers of the selected provider. However, to calculate the vectors, all the data that you index must be uploaded to the cloud. Of course, we don't know what the provider does with your data after the calculation, but we find it highly unlikely that it will simply be deleted.

There are certainly others besides OpenAI from whom you can rent embedding models: Google, MistralAI, CohereAI, IBM, Oracle, Nvidia, and so on—the list reads like a who's who of the IT industry. Examples of use with the LlamaIndex library can be found at *https://docs.llamaindex.ai/en/stable/examples* under **Embeddings**.

You should also keep in mind that the embedding model isn't only used to create the index but also for every question you ask. As we'll see in Section 12.6, the same embedding model that was used to create the index must also be used to convert your questions into vectors.

12.5   Vector Store Databases

To avoid having to recalculate the vector embeddings every time the application is started, we'll save the index in a database. The task of the vector database is to store the numerical representation of unstructured data (the vectors with their dimensions) and to search for them using various algorithms. The implementations are divided into those in which external libraries perform this search in an existing (relational or NoSQL) database, those in which plug-ins are used in the database, and *genuine* vector databases.

Real vector databases are on a par with SQL and NoSQL databases. While SQL databases force the source data into a strong structure (through rigid tables with many data types), NoSQL structures the data only roughly through key-value pairs with data types such as numbers, strings, or dates. With vector databases, the source data is stored unstructured, and only the calculated vectors are used in the search.

A prominent representative of genuine vector databases is Milvus. The database is available under the free Apache 2 license and can be used very easily in different scenarios: The Lite version is ideal for *prototyping* or on devices with very low-end hardware. For production deployments, *Milvus Standalone* can be started very easily in the Docker container. If you think big, you can also start Milvus in a Kubernetes cluster in a cloud-native architecture. In addition, you can also rent Milvus as a ready-made cloud solution from Zilliz, the company behind the development.

We tried out both the Lite version and the standalone version. Both variants were convincing in our very small dataset. The Docker setup with a compose.yaml file can be started practically at the touch of a button. The API calls don't differ whether you use Lite/standalone or the cloud version of Milvus. This allows you to turn your small test project into a scalable solution in just a few seconds.

We use the standalone version of Milvus in the Docker setup with compose for our RAG project. If you've installed the latest version of Docker and the compose plug-in, all you need to do is download the configuration file provided on GitHub and start the setup via docker compose up -d:

https://github.com/milvus-io/milvus/releases/download/v2.4.6/milvus-standalone-docker-compose.yml

In addition to the database, two other services are started: the *etcd* configuration service and the *MinIO* object store. The database becomes available shortly afterward on port 19530 without authentication.

Zilliz has also developed a graphical user interface (GUI) for managing the database as open-source software showing the indexed PDF documents and web pages. On the left of the screen are three collections, each created with different embedding models (see Figure 12.3).

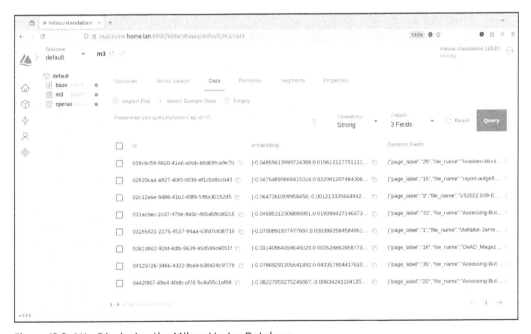

Figure 12.3 Attu Displaying the Milvus Vector Database

Attu can easily be added as an additional container to the Docker Compose setup. Simply add the following lines to the `services` section of the downloaded Compose file:

```
attu:
  image: zilliz/attu:v2.4
  environment:
    MILVUS_URL: milvus-standalone:19530
  ports:
    - "8000:3000"
```

After starting the container using `docker compose up -d`, you can access the web interface at *http://localhost:8000* or the host name on which you started the docker compose setup. We found the insight into the internal data structure of the vector database very fascinating.

To use the Milvus database for our sample project, we include a `StorageContext` LlamaIndex when creating the `VectorStoreIndex`:

```python
from llama_index.vector_stores.milvus import MilvusVectorStore
from llama_index.embeddings.ollama import OllamaEmbedding
from llama_index.core import StorageContext
vector_store = MilvusVectorStore(uri="http://localhost:19530",
                                 dim=1024, overwrite=True)
storage_context = StorageContext.from_defaults(
  vector_store=vector_store
)
index = VectorStoreIndex.from_documents(
    documents=documents,
    transformations=[
        SentenceSplitter(),
        TitleExtractor(),
        OllamaEmbedding(model_name="mxbai-embed-large"),
    ],
    storage_context=storage_context,
    show_progress=True,
)
```

In addition to `uri`, which refers to the database previously started in Docker, the dimension of the vectors must also be specified and must then be adapted to the embedding model used. The `mxbai-embed-large` model used here uses vectors with 1,024 dimensions. To avoid duplicates, the `overwrite=True` parameter deletes the database before the index is created. The application is now ready, and we can try out the first queries in natural language using the indexed content.

12.6   RAG Queries

For our question to lead to a meaningful answer, several steps are executed in succession by the LlamaIndex library:

1. The question in natural language is converted into vectors. For this purpose, the same embedding model must be used as when creating `VectorStoreIndex`.
2. The system searches for similar vectors in the database.
3. The content of the LlamaIndex nodes found is redirected to an LLM as a context with the question.
4. The LLM formulates an answer in natural language from the data.

Having dealt with the embedding model so far, this is the first time we've come into contact with the LLM. But that's not quite true: in Section 12.4, `TitleExtractor` used an LLM to create a summary from the split documents. Which LLM is used, and where can we configure it?

LlamaIndex currently uses OpenAI's `gpt-4o` as the default LLM, provided that you've stored a valid API key in the `OPENAI_API_KEY` environment variable. However, you can change this behavior very easily and use a local `llama3.1` from your Ollama server, for example:

```
from llama_index.core.settings import Settings
from llama_index.llms.ollama import Ollama
...
Settings.llm = Ollama(model="llama3.1")
```

Alternatively, you can also use the popular Claude 3.5 Sonnet:

```
from llama_index.llms.anthropic import Anthropic
...
Settings.llm = Anthropic(model="claude-3-5-sonnet-20240620")
```

Note the import of `settings` and the subsequent `Settings.llm` setting. LlamaIndex now uses the set model for all accesses to an LLM. The same applies to the embedding model. The setting for this is `Settings.embed_model`.

> **Prompt:** Which butterfly species was found most frequently in Tyrol in 2023?

To ask this question, we still need a LlamaIndex `query_engine`, which we can derive from the `VectorStoreIndex`:

```
query_engine = index.as_query_engine()
response = query_engine.query(
  "Which butterfly species was found "
  "most frequently in Tyrol in 2023?")
print(response)
```

The different answers from the various models are fascinating. The preceding question refers to a table on page 41 of the 2023 annual report on insect monitoring in Austria. The PDF document contains a wide range of information on the scientific results, not only on the investigations in Tyrol. Here is the answer from the combination of the bge-m3 embedding model and the Claude 3.5 Sonnet LLM:

> The large ox-eye butterfly (Maniola jurtina) was observed most frequently in Tyrol in 2023. With 430 individuals counted, it was the species with the most sightings that year.

The information is even more detailed when we use the OpenAI embedding model:

> The large ox-eye butterfly (Maniola jurtina) was the butterfly species most frequently observed in Tyrol in 2023. This species remains at the top of the list with the most individuals counted. This is followed in frequency by the small cabbage white butterfly (Pieris rapae), the common blue butterfly (Polyommatus icarus), and the small meadow bird (Coenonympha pamphilus).

> However, it's important to note that the small fox (Aglais urticae) was the most widespread species and was detected at 87 out of 100 sites surveyed.

If you've ever wanted to extract data from a table in a PDF document yourself, then you'll know that in most cases, it's faster to enter the data again. PDF describes where something is on a page, but not how it belongs together in terms of content. This makes it all the more astonishing how well the combination of vector search and LLM works. The fact that Claude points out that the small fox was found at most locations is remarkable, even if the number 87 refers to the entire period between 2018 and 2023 and not to the sought-after year 2023.

You can download the PDF document from the project website to get your own impression of the initial data at *https://viel-falter.at/ergebnisse/veroeffentlichungen*.

12.6.1 Source Code

After all the code snippets, we want to show you the complete program that we used to test the models. It contains command-line parameters for selecting the embedding model and the LLM as well as a switch to regenerate the VectorStoreIndex and a switch for more logging output. We haven't printed the many `import` statements as they aren't necessary to understand the program.

Note that when creating the `VectorStoreIndex`, the value of the `args.embed` switch is transferred as the name of the *collection*. If this switch isn't specified, the `openai` default value will be used. This allows you to save several indexes in parallel in the Milvus database and test them in parallel.

One more note on the environment variables for OpenAI and Anthropic: we've created an `.env` file in the current directory with the following content:

```
OPENAI_API_KEY="sk-proj-dG...."
ANTHROPIC_API_KEY='sk-ant-api03-....'
```

The Python library `dotenv`, which you can install using `pip install dotenv`, loads this file. Provided you still have some credit left in your accounts, you can try out the various models from OpenAI or Anthropic.

```
# File: vf-rag.py
load_dotenv()
parser = argparse.ArgumentParser(
    prog="vf-rag",
    description="Information about viel-falter.at project")
parser.add_argument("-v", "--verbose", action="store_true")
parser.add_argument("-d", "--debug", action="store_true")
parser.add_argument("-r", "--regenerate", action="store_true")
parser.add_argument(
    "-l", "--llm", choices=["ollama", "openai", "claude"],
    default="claude")
parser.add_argument(
    "-e", "--embed", default="openai",
    choices=["m3", "nomic", "mxbai", "base", "openai"])
args = parser.parse_args()

if args.verbose or args.debug:
    logging.basicConfig(
        stream=sys.stdout,
        level=logging.DEBUG if args.debug else logging.INFO)
    logging.getLogger().addHandler(logging.StreamHandler(
      stream=sys.stdout))

milvus_store = "http://localhost:19530"
dimensions = 1024
if args.embed == "m3":
    Settings.embed_model = HuggingFaceEmbedding(
        model_name="BAAI/bge-m3",
```

```
            cache_folder="/home/user/huggingface_cache")
elif args.embed == "base":
    Settings.embed_model = HuggingFaceEmbedding(
        model_name="BAAI/bge-base-en-v1.5",
        cache_folder="/home/user/huggingface_cache")
    dimensions = 768
elif args.embed == "nomic":
    Settings.embed_model = OllamaEmbedding(
    model_name="nomic-embed-text")
    dimensions = 512
elif args.embed == "mxbai":
    Settings.embed_model = OllamaEmbedding(
    model_name="mxbai-embed-large")
else:
    Settings.embed_model = OpenAIEmbedding()
    dimensions = 1536


if args.llm == "ollama":
    Settings.llm = Ollama(model="llama3.1")
elif args.llm == "openai":
    Settings.llm = OpenAI(model="gpt-4o-mini")
else:
    Settings.llm = Anthropic(
      model="claude-3-5-sonnet-20240620")


logging.log(level=logging.INFO, msg=args)
logging.log(level=logging.DEBUG, msg=os.environ)
if args.regenerate:
    pdf_documents = SimpleDirectoryReader("pdfs").load_data(
      show_progress=True)
    webpages = [
        "https://viel-falter.at/",
        "https://viel-falter.at/monitoring/ziele-vision/",
        "https://viel-falter.at/projektpartner/",
        "https://viel-falter.at/news/",
        "https://viel-falter.at/tagfalter-exkursion/",
    ]
    web_documents = SimpleWebPageReader(
      html_to_text=True).load_data(webpages)
    documents = pdf_documents + web_documents

    vector_store = MilvusVectorStore(
        uri=milvus_store, dim=dimensions, overwrite=True,
        collection_name=args.embed)
```

```
    storage_context = StorageContext.from_defaults(
        vector_store=vector_store)
    index = VectorStoreIndex.from_documents(
        documents=documents,
        storage_context=storage_context,
        show_progress=True)
else:
    vector_store = MilvusVectorStore(
        uri=milvus_store, dim=dimensions,
        collection_name=args.embed)
    index = VectorStoreIndex.from_vector_store(
        vector_store)

query_engine = index.as_query_engine()
response = query_engine.query(
  "Which butterfly species was found "
  "most frequently in Tyrol in 2023?")
print(response)
```

12.6.2 Conclusion

We were impressed by how much can be achieved using the LlamaIndex library, even with the default settings. The entire source code of the RAG script just presented comprises only 100 lines of code, whereby we've also included command-line parameters for the LLM to be used and the new creation of the index for our tests.

We don't want you to believe that you can create an application with such simple tricks that can be used successfully in a production environment. There are several examples of how a chatbot that was quickly put together wasn't a great success. Problems with LLMs such as *bias* can't be ignored. In addition, the application should also be secured against *creative* user input that attempts to use the LLM for purposes other than the queries you require.

For a company-internal knowledge database, however, this could be a starting point for a successful application. There is no question that tests should be run continuously to check the quality of the answers according to previously defined rules.

12.7 Text-to-SQL

The internet is teeming with examples that use an LLM to filter a database with one table and five columns to the largest value ("Which department generated the most sales?"). It's hardly surprising that an LLM to which the table structure is passed as a context can formulate a corresponding SQL query. With real-life data, things look a little different.

In the following sections, we'll look at the dataset of the viel-falter project. In this research project, which we've already used for content in the first part of this chapter, butterflies are identified according to their species in designated areas.

12.7.1   Database Export from MongoDB to SQLite

The data for the project has been stored in MongoDB, a NoSQL database, for more than 10 years. At MongoDB, efforts have been underway for some time to make the database content searchable using AI technology, but this mainly concerns the cloud offerings (MongoDB Atlas) and not the open-source version. For our example, we wanted to test the capabilities of current LLMs in dealing with SQL databases.

As NoSQL databases have structures such as arrays or nested documents, exporting to an SQL database isn't easy. However, with the knowledge of which data is to be exported to which tables, we (authors and AI) can create an export script that takes over this function.

We've printed excerpts of a sample document from the MongoDB surveys collection here:

```
{
  "_id": {
    "$oid": "669e162e2a7d5ed3d645257d"
  },
  "start": "13:52",
  "end": "14:28",
  "site": {
    "_id": {
      "$oid": "5e60d1d095c061fecc14670f"
    },
    "name": "132_alm_1",
    "community": "Lech",
    "nuts2": "Vorarlberg",
    "stratum": "alm",
    "climaticRegion": "A",
    "geometry": {
      "type": "Point",
      "coordinates": [10.159735679626467, 47.21806151908246]
    }
  },
  "notes": "Maculinea arion ausserhalb Erhebung",
  "speciesList": [
    {
      "id": "595df3ce912bcbcc8f70765d",
      "count": 2
```

```
    },
    {
      "id": "595df3ce912bcbcc8f70769c",
      "count": 1
    }
  ],
  "lastMod": {
    "$date": "2024-07-22T08:19:58.232Z"
  },
  "date": {
    "$date": "2024-07-21T11:52:00.000Z"
  }
  ...
}
```

The final prompt we used to make Claude 3.5 Sonnet export parts of the MongoDB database to an SQLite database was as follows:

> **Prompt:** Write a Python script to export a MongoDB collection surveys from a database vielfalter to SQLite. Extract the array field speciesList to a separate table speciesList. Note that this is an array of documents, with an id and count field.
>
> Extract the embedded documents in site to a separate table, including fields _id, name, community, and nuts2, removing duplicates and linking it to surveys. Include fields _id, date, start, end, and notes for the surveys table.
>
> Export the collection species to a table including fields _id, german, and latin.

The resulting Python script was executable without modification, connected to the locally running copy of the database and exported the data to an SQLite database.

However, we spent several hours tweaking this prompt. Originally, we wanted to create a generic export script that would automatically recognize the arrays and embedded documents and create the SQLite tables accordingly. However, we were unable to wrap these requirements into a single prompt.

What worked surprisingly well, however, was an iterative approach in which we further developed the script in a chat with the LLM. With a little manual work at the end (we simply ran out of patience for the dialogs with the AI tool), we came to a usable result with the following prompts.

> **Prompt:** How do I export a mongo collection to CSV where arrays in the collection get separate CSV files with foreign keys?

Interestingly, our results were better when we exported to comma-separated values (CSV) first and brought SQLite into play later. Unfortunately, the first script we received in response to this prompt wasn't executable. Claude pointed out the error and was able to rectify it.

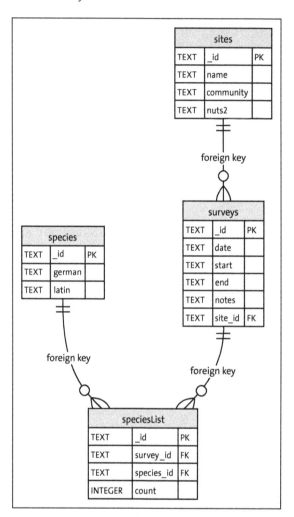

Figure 12.4 The Entity-Relationship Diagram of the Exported SQLite Database

Prompt: Python returns an error for this script:

```
Traceback (most recent call last):
  File "/data/export.py", line 31, in <module>
    writer.writerow(['parent_id'] +
      first_doc[array_field][0].keys())
TypeError: can only concatenate list (not "dict_keys")
to list
```

The CSV export worked, but we had JSON strings in the exported CSV files, which were originally arrays of documents in our collection.

Prompt: Great! Can you improve a little bit, so that array fields aren't JSON strings but separate columns in CSV files?

Claude did this with flying colors. The field names of the documents were recognized and created as columns in CSV. Next, we wanted to store the embedded document for sites in a separate CSV file, as is logical in an SQL database according to the normal forms.

Prompt: Perfect. What would I need to do if I want to normalize one key of the document, where the value holds an embedded document?

As the embedded documents often describe the same sites, we wanted to avoid duplicates.

Prompt: Great. Can you improve a bit more, so that we don't get duplicate entries in the new CSV file with the embedded document?

Claude was also able to solve this task. Although the embedded MongoDB ID wasn't used for comparison, but a hash was created using the document, the approach also works. Now we wanted to create an SQLite database instead of CSV files.

Prompt: Thanks, this works perfectly. Can you alter the script to write the data to an SQLite database instead of CSV files?

And lo and behold, Claude was able to correctly add the necessary libraries to the script and write the export to an SQLite file. During the many tests, we always had to adjust the collection name for each test. To avoid this, we instructed the LLM to use the correct name in the script. We also needed the species collection in the SQLite database, which contains the list of all butterflies, just like in a relational database.

Prompt: Excellent. Can you rename the exported collection main_collection to table surveys and export another collection named species from the same Mongo database to the same SQLite database?

The resulting Python code, which we haven't printed here, is quite readable, even if a Python expert would probably have designed one or two functions more efficiently. The result was sufficient for our purposes. Would we have been faster if we had written the export script ourselves? We can answer this question in the affirmative.

12.7.2 Text-to-SQL with LlamaIndex

We can make the exported SQLite database available to the LlamaIndex library without hesitation because the production data remains in the MongoDB, to which the script has no access. Of course, you can also connect a productive SQL database here instead of an export. The SQLAlchemy library we used supports common relational databases such as MySQL/MariaDB, PostgreSQL, Microsoft SQL Server, and Oracle. The problem is that you don't know in advance which SQL statements the AI tool will create. Even if the AI tool only has read-only access to the tables, a database administrator will probably roll their eyes at the prospect of uncontrolled SQL queries.

The LlamaIndex library makes Text-to-SQL queries child's play. The NLSQLTableQuery-Engine module generates the *QueryEngine* that we use for queries in natural language. An *SQLAlchemy engine* with the reference to the database file is created beforehand. You must also install this Python module using the pip install SQLAlchemy command:

```
from llama_index.core import (
    Settings,
    SQLDatabase,
)
from llama_index.core.query_engine import NLSQLTableQueryEngine
from llama_index.llms.anthropic import Anthropic
from sqlalchemy import create_engine

tables = ["surveys", "speciesList", "species", "sites"]
engine = create_engine("sqlite:data/vielfalter.db")
sql_database = SQLDatabase(
    engine,
    include_tables=tables
)
query_engine = NLSQLTableQueryEngine(
    sql_database=sql_database,
    tables=tables
)
Settings.llm = Anthropic(model="claude-3-5-sonnet-20240620")
response = query_engine.query(
  "How many species were found in 2024?"
)
print(response)
```

To answer the final question about the number of different species found in 2024, the LlamaIndex library first reads the structure of the listed tables. As you can imagine, the naming of tables and columns is very important so that the LLM can later generate correct SQL commands.

```
INFO:llama_index.core.indices.struct_store.sql_retriever:>
  Table desc str: Table 'surveys' has columns: _id (TEXT),
    date (TEXT), start (TEXT), end (TEXT), notes (TEXT),
    site_id (TEXT),
    and foreign keys: ['site_id'] -> sites.['_id'].

Table 'speciesList' has columns: _id (TEXT), survey_id (TEXT),
  species_id (TEXT), count (INTEGER),
    and foreign keys: ['survey_id'] -> surveys.['_id'],
    ['species_id'] -> species.['_id'].
...
```

This information is sent to the LLM together with detailed instructions for creating the SQL query (we'll come to this in a moment), which in turn sends back an initial SQL suggestion:

```
Predicted SQL query:
SELECT COUNT(DISTINCT sl.species_id) AS species_count
  FROM surveys s
  JOIN speciesList sl ON s.`_id` = sl.survey_id
  WHERE strftime('%Y', s.date) = '2024'
```

The LlamaIndex library then connects to the SQLite database and executes the query. The result gets included in a new prompt and is resent to the LLM as SQL Response: [(79,)]. The LLM formulates the short and correct answer (at the time of writing):

> Based on the SQL query results, 79 distinct species were found in surveys conducted in 2024.

Even if this sounds like a very simple question, Claude has to demonstrate a bit of SQL know-how. The speciesList table must be linked to the surveys table and the date field must be analyzed correctly. The fact that we've formulated the prompt in English helps the LLM create the SQL query. In our tests, the large LLMs also succeeded in converting German-language prompts into correct SQL with the English table structure, but the success rate for complex queries was significantly lower.

But that was just the beginning. In the next section, we'll ask more difficult questions and have them answered by different language models.

12.7.3   Sample Queries

Our program starts quite quickly (in Python, "quickly" always leaves room for interpretation), but we don't want to restart it for every query. Using a while loop in the program, we constantly request new prompts:

```
from prompt_toolkit import PromptSession

session = PromptSession()
try:
    while True:
        q = session.prompt("Query our viel-falter database: ")
        if q == "exit":
            break
        answer(q)
except (KeyboardInterrupt, EOFError):
    print("bye")
```

The `PromptSession` function from the `prompt_toolkit` (`pip install prompt_toolkit`) saves a list of queries that have already been submitted. You can scroll through older entries using the `Ctrl`+`P` shortcut and even search in these entries via `Ctrl`+`R`. The program can be terminated by entering `exit` or using `Ctrl`+`C` (`KeyboardInterrupt`) or `Ctrl`+`D` (`EOFError`). Now we want the AI tool to answer a few more difficult questions.

> **Prompt:** How many surveys were conducted in 2024 and on how many different sites?

> Based on the SQL query results, in 2024:
>
> 1. A total of 586 surveys were conducted.
> 2. These surveys were carried out across 184 different sites.

The answer sounds good, but can we believe it? It would be interesting to see the SQL query that led to this result. Fortunately, we don't have to do much additional programming, but only activate logging for the LlamaIndex library. We use the same command-line parameters as in the previous example (Section 12.6) to activate logging. The SQL commands are displayed if the log level is set to DEBUG. To do this, we start the application with the `--debug` switch.

The log messages contain detailed logs of the connections to the LLM (we've chosen Claude 3.5 Sonnet) and the prompts that LlamaIndex generates and sends to the LLM. The rather detailed prompt that is used to generate the SQL query is interesting:

> **LlamaIndex-Prompt:** Given an input question, first create a syntactically correct SQLite query to run, then look at the results of the query and return the answer. You can order the results by a relevant column to return the most interesting examples in the database. Never query for all the columns from a specific table, only ask for a few relevant columns given the question. Pay attention to using only the column names that you can see in the schema description. Be careful not to query for columns that don't exist.

Pay attention to which column is in which table. Also, qualify column names with the table name when needed. You are required to use the following format, each taking one line:

Question: Question here

SQLQuery: SQL Query to run

SQLResult: Result of the SQLQuery Answer

The prompt is then extended by the structure of all tables we've specified in the program (the `tables` variable) and, of course, by the question asked. In the further course of the communication, we see the suggested SQL query:

```
Predicted SQL query:

SELECT COUNT(*) AS total_surveys,
       COUNT(DISTINCT site_id) AS distinct_sites
  FROM surveys WHERE date LIKE '2024%'
```

A date comparison using `LIKE` may seem a little unconventional for SQL purists, but in SQLite, the query works as desired. We find the answer in the log output as `SQL response: [(586, 184)]`. The nicely formulated answer then generates the LLM again, which formulates English sentences from the numbers and the context. Thus, Claude Sonnet is doing a pretty good job. Now we wanted to find out how the local models in an Ollama server fare in this regard. Meta's current open-source model, `llama3.1`, initially gave us little hope:

> Based on the query response, it appears that there were no surveys conducted in 2024.
>
> Therefore, the synthesized response would be:
>
> "No surveys were conducted in 2024. Additionally, because there were no surveys, there are also no different sites associated with these surveys."

The analysis of the SQL query immediately revealed the problem:

```
SELECT COUNT(DISTINCT site_id) AS total_sites,
       COUNT(*) AS total_surveys
  FROM surveys WHERE date = '2024';
```

Most of the query was actually correct, only the date field wasn't analyzed correctly, which is disappointing. We immediately tried the query again and were somewhat surprised that `llama3.1` suddenly had a solution ready, but with a different result than the AI from Anthropic (Claude) before. (Llama 3.1 carried out the highlighting in the result.)

> According to our records, **76 different sites** were surveyed in **2024**, with a total of **500 surveys** conducted across these locations.

Debug logging helped us solve this mystery as well. The AI tool had tried to do too much here and linked the list of species found (speciesList) to the surveys table. But there are also surveys in which no butterflies are found, and that is also a result (the fact that the differences are so great is also due to the fact that not all surveys are carried out at species level, but that would go too far here).

```
Predicted SQL query:
SELECT COUNT(DISTINCT T1.site_id) AS num_sites,
    COUNT(T1._id) AS num_surveys
  FROM surveys AS T1
    INNER JOIN speciesList AS T2 ON T1._id = T2.survey_id
  WHERE STRFTIME('%Y', T1.date) = '2024'
```

The story got a little scary when we ran the query again and got the right result, which the Claude AI tool had provided straight away. The three tests took place within one minute, and the Ollama server had no other computing tasks. Perhaps the llama3.1 model isn't good enough for this task. We're also testing the deepseek-coder-v2 model in Ollama. However, our initial confidence was dampened again with more complex queries, as we were faced with the same problem of answers being incorrect and different for each query.

Our impression was that these smaller local models can't handle complex Text-to-SQL tasks properly. By the time you hold this book in your hands, these models will certainly be much more advanced and may be able to execute Text-to-SQL well locally.

Claude AI and OpenAI's gpt-4o, on the other hand, have repeatedly surprised us. In the following query, how many individuals of a specific species were found in a region in one year, all four tables of our SQLite database had to be joined in the SQL query, and the numeric count field in one of the tables had to be summed up. The nuts2 database field is derived from the European classification of territorial units (*https://ec.europa.eu/eurostat/web/nuts*).

Prompt: How many individuals of species "Admiral" were found in surveys in 2024 in the Nuts2 region of Vorarlberg?

Based on the SQL query results, 4 individuals of the species "Admiral" were found in surveys conducted in 2024 in the NUTS2 region of Vorarlberg.

We couldn't have written the SQL query used for this much better ourselves:

```
SELECT SUM(sl.count) as total_count
  FROM speciesList sl
  JOIN surveys s ON sl.survey_id = s._id
  JOIN species sp ON sl.species_id = sp._id
  JOIN sites st ON s.site_id = st._id
```

```
WHERE sp.german = 'Admiral'
  AND strftime('%Y', s.date) = '2024'
  AND st.nuts2 = 'Vorarlberg'
```

By playing around with the database, we also found incorrect entries by chance. The search for all existing Nuts2 regions led to a surprising result.

Prompt: How many different Nuts2 regions are in the database and list them?

Based on the SQL query results, there are 11 different Nuts2 regions in the database. The regions are:

1. Tirol
2. Vorarlberg
3. Südtirol
4. Niederösterreich
5. Salzburg
6. Oberösterreich
7. Kärnten
8. Steiermark
9. Burgenland
10. Wien

It's worth noting that there appears to be an empty entry in the list (indicated by the double comma between Burgenland and Wien), which could represent a null or empty value in the database. This empty entry is counted as one of the 11 distinct regions, even though it doesn't have a name.

However, the AI tool immediately explains where the 11th entry, which doesn't appear in the list, comes from: there is one entry where the region is obviously empty. Without a little knowledge of the database structure, it will be difficult to find the incorrect entry. The SQL query that led to the result also helps us here:

```
SELECT COUNT(DISTINCT nuts2) AS count,
       GROUP_CONCAT(DISTINCT nuts2) AS regions
FROM sites
```

This is an entry in the `sites` table. Now we can ask the AI tool which entry in this table it is:

Prompt: Show me sites not assigned to a Nuts2 region. Include database IDs for these sites.

Based on the query results, there is one site that isn't assigned to a NUTS2 region. The details are as follows:

Database ID: 65f6a6f616634be52c735711

Name: [No name provided]

It's worth noting that while the database ID is present, the name field appears to be empty for this site. This could mean either the site doesn't have a name recorded in the database, or there might be an issue with the data entry for this particular site.

The incorrect entry was quickly corrected using the database ID.

12.7.4 Conclusion

The Text-to-SQL concept has great potential for us. This concept could save database administrators work, especially in environments where people without a technical background want to sporadically request different key figures from a central database. Will this make the database admins superfluous? Certainly not at the moment.

The problem of not being able to rely on the AI tool's answers, even with Text-to-SQL, makes the system unsuitable for a business-critical area. But the same also applies to hallucinating models for other tasks.

Another limiting factor is currently the number of tokens an LLM can process in context. If the SQL queries contain too much data, the LLM can no longer process the responses. However, we also expect rapid development in this field. It will definitely be exciting to see where this technology can establish itself in the future.

Chapter 13
Risks and Outlook

We already addressed this question in the foreword: is the glass half full or half empty? AI tools undoubtedly help to increase your efficiency as a developer or administrator, as a programmer, or as a database designer. However, it's equally undisputed that AI tools create errors, sometimes even major ones.

The errors are all the more serious because current AI tools lack inherent self-control or self-criticism. Instead of a simple "I don't know," you'll receive a polite answer that is elegantly formulated and convincingly argued, but nevertheless completely wrong.

In this chapter, we take a look at the risks and problems of AI tools:

- First, we provide an overview of the basic risks: outdated code, poor maintainability, data protection concerns, and so on.

- In Section 13.2, we provide concrete examples in which AI tools have returned clearly incorrect answers. Many examples have been taken from our own work. We also cite some studies on this topic.

- Is it even ethically justifiable to use AI tools? In our view, there is no objectively clear answer to this question. However, we're trying to explore the ethical and, in some cases, legal aspects of using AI tools for programming.

- Finally, we summarize where AI-supported coding currently stands and where we think the journey is heading. To get one point out of the way, we're convinced that the use of AI tools is only responsible within the scope of one's own knowledge. AI tools are therefore no substitute for solid IT training.

13.1 Issues and Limitations of Using AI Tools

While working on this book and while programming for other projects, we encountered various issues when using AI coding aids. Accordingly, we've repeatedly referred to limitations in the preceding chapters. At this point, we want to summarize once again where we believe the biggest limitations and pitfalls can be found when using AI tools.

13.1.1   Old, Cumbersome Code

Any language model is only as good as its training material. For software development in particular, training suffers from the fact that there is always more old material available on the internet than up-to-date material. There is an infinite number of pages that refer to old Java versions, but only very few pages in which new functions added to Java are used in a way that makes sense. (You can replace Java with the programming language of your choice in this and the following paragraphs.)

ChatGPT therefore tends to suggest code that uses old, established functions or libraries. That code will usually work. However, if you use a current version of Java, there might have been a new, more elegant, and more efficient way to solve your problem. You can try to specify the version number you use in the prompt (say: Java 21), but even that only helps to a limited extent.

13.1.2   The Question of Maintainability

AI tools enable you to produce lots of lines of code in no time. Your productivity measured in this way will be great, and your boss will be happy about you. But who maintains this code? Who fixes a bug that only occurs under very specific circumstances and is discovered for the first time after three months?

We've already shown in this book that ChatGPT and others can also be a great help in troubleshooting. But what is even better is compact, well-designed code that works right from the start.

Note that by no means do we deny that AI tools often provide excellent code. Unfortunately, this isn't always the case. An experienced developer will probably scrutinize the AI code and perhaps make a second attempt. But many programmers will be pragmatic and carry out a short test to see whether the new function does what it's supposed to do and leave it at that—even if, objectively speaking, the code isn't ideal.

13.1.3   Privacy

If you use free AI services from commercial providers (OpenAI, Microsoft, Google, etc.), your prompts, the resulting responses, and your reaction to them will be stored and used to train future language models. So far, so fair. As with many other services that are free, you pay with your data.

For many companies, however, code is a company secret. Self-developed algorithms make up part of the company's value. Such companies don't feel comfortable giving editors with AI support virtually unhindered access to the entire code.

Of course, AI company offerings (e.g., GitHub Copilot Business) promise a certain level of privacy. However, it's difficult to assess, let alone verify, how far this goes and whether the rather vague promises ("your data is excluded from training by default") are actually kept.

Even if AI companies try to comply with privacy regulations to the best of their knowledge and belief, mistakes do happen. AI tools are slapped together; whoever is first on the market with good AI products can gain significant market share. Quality control and code reviewing can sometimes fall by the wayside.

Moreover, since Edward Snowden's revelations, it's not a conspiracy theory but a fact that the US intelligence services have almost unrestricted access to US cloud storage. AI companies that operate almost exclusively in the United States are hardly exempt from this.

Keep Things in Perspective

Do you store your code in GitHub repositories? Do you use Visual Studio Code (VS Code) with all conceivable plug-ins? Then, your code is already doubly compromised. The use of a commercial coding wizard doesn't dramatically worsen the situation.

Conversely, if you're reluctant to use AI tools for coding due to privacy considerations, you also need to question your other security measures. You should definitely run your Git repositories yourself (e.g., in a GitLab instance on a company server). You should also be very selective when choosing development tools and the associated plug-ins. The editor must be able to read your code, as must most plug-ins. A malicious plug-in can therefore compromise your code.

13.1.4   A New Kind of Dependency

We don't know how often or how actively you cycle. You have the choice of relying solely on your own muscle power or being assisted by a motor. The first ride on an e-bike is definitely a pleasure! Regardless of whether you want to conquer a small hill or an entire mountain pass—you'll be much faster and feel more comfortable.

It doesn't take long for you to get used to the support provided by the motor and battery. "Ordinary" cycling is no longer fun at first, and a year or two later, you'll also lack muscle power. There's no turning back, and the temptation to use turbo mode (or whatever the highest support level on your e-bike is called) more and more often becomes ever greater.

You can probably guess what this comparison is aimed at. Programming with AI support isn't only more efficient, it's also more convenient. Over time, you no longer know the syntax of frequent constructions by heart, no longer think when formulating loops, and begin to rely more and more on the code provided by your AI tool. You'll run a quick test to see if everything works as expected and then it's on to the next step!

It doesn't take long before you become dependent on AI tools. Even if you haven't (completely) forgotten how to program "properly," without AI support you won't be as agile and efficient as you used to be.

This can be particularly fatal if political, economic, or legal reasons mean that AI services can no longer be offered in a country. Imagine a court ruling that AI services from OpenAI, Microsoft, Apple, or another IT company don't comply with EU data protection regulations. To avoid a long dispute or high payments, the service will simply get switched off in the EU. That would be a rather unpleasant prospect, also for competitive reasons. (How are European developers without AI support supposed to keep up with American programmers using these tools?)

13.1.5   Learning to Program in the Age of AI

Basically, this book is intended for readers who already know how to program. But let's take a very brief look at the next generation. Learning new skills requires that you actually practice them. Writing essays is certainly a tedious exercise at school, but it's essential if you want to be able to formulate convincing texts yourself later on. Similarly, the initially frustrating compilation of your first lines of code is important for learning basic programming techniques.

Finding a responsible scale for the use of AI tools is currently a huge challenge for schools and universities. To make matters worse, there is a lack of suitable control mechanisms. Self-programmed code is best recognized by the errors it contains. But it's of course inadmissible to immediately place every usable code under general suspicion: "But you did this exercise with AI help, didn't you?"

13.2   Exemplary AI Failure

If you use a coding wizard such as GitHub Copilot, the AI tool constantly makes code suggestions. If you type for, the wizard suggests a loop over a data structure that appears in the previous lines. Entering // or # gets completed to a full comment by the AI tool, and so on. It's clear that only a fraction of these unsolicited suggestions are useful.

It's up to you to ignore unsuitable suggestions and avoid being distracted if possible. In this section, we don't consider inappropriate suggestions to be errors; we don't want to accuse the AI of not being clairvoyant. (You often get this impression, but it's deceptive.) This section deals with "real errors" in which AI tools provide incorrect or misleading answers despite clear prompts.

13.2.1   Outdated Information

When we were writing this book, questions about the OpenAI API were the most reliable way to go straight to errors. OpenAI has fundamentally changed its API after the cut-off for the training of ChatGPT 4o. This led to the absurd situation that ChatGPT was unable to provide correct information on the use of the parent company's API. (Most other language models were also overwhelmed because it usually takes several months between the start of training for a language model and its publication. By the time you read this book, the situation will almost certainly have improved—unless the API changes again, that is.)

> **Prompt:** I need a Python script to use the OpenAI API as a chat system for programming. The output should be formatted to be readable in a terminal (colors are OK, but no Markdown syntax). In addition, the language model should favor short answers (2 paragraphs max) and code only if possible.

> Certainly! Here's a Python script to use the OpenAI API as a chat system for programming assistance. The output will be formatted for terminal readability, with colors for clarity.
>
> First, ensure you have the `openai` and `termcolor` libraries installed:
>
> `pip install openai termcolor`
>
> Then, you can use the following script:
>
> [code and more details]

We've tried the code, but it doesn't work. The `openai` module doesn't recognize the classes and methods suggested by ChatGPT because they were restructured and renamed after the model was trained. We draw ChatGPT's attention to the error.

> **Prompt:** The code doesn't work, it's not compatible to 1.0 API. Can you help?

> Sure, let's update the script to be compatible with OpenAI's latest API version, which uses the `openai.ChatCompletion.create` endpoint.
>
> [code and more details]

The new code doesn't work any better than the one in the first answer. In a way, you can't blame the AI tool here. The training material was compiled in October 2023. One month later, the OpenAI API was fundamentally changed as part of the 1.0 release. Ultimately, the timing in this example was simply bad luck.

> **Prompt:** What is your training cut-off date?

> My training data includes information up until October 2023. Any events or developments that have occurred after that date aren't within my knowledge base.

Well, we could point ChatGPT to the current documentation:

> **Prompt:** I need a Python script to use the OpenAI API as a chat system…
>
> Please note that the API has changed since your training. Up-to-date documentation is available here:
>
> *https://platform.openai.com/docs/guides/text-generation*

Sure, here is a Python script that uses the OpenAI API to create a chat system for programming, with a preference for short answers and code only.

[code and more details]

The code remains incorrect even at the third attempt. What is remarkable about the example is the confidence with which ChatGPT introduces false answers ("Certainly," "Sure," etc.). At the last prompt, there is no mention in the reply that the link to the current documentation was simply ignored.

13.2.2   Hallucinations

When language models answer questions for which there is no reliable information, they sometimes invent answers. In the context of program development, AI tools suggest the use of modules, libraries, or functions that don't even exist.

Hallucinations occur because language models use statistical procedures to generate the answer; that is, they continue a text as plausibly as possible once it has been started. It can happen, for example, that a function available in programming language X appears in the answer that actually refers to language Y. Common language models are so "smart" that they adapt the function name to the conventions of the respective programming language (in PowerShell, e.g., to the verb-noun terminology, such as `Get-ClusterNameObject`).

Hallucinations happen most frequently in peripheral IT areas where there is little training material. As long as you use AI tools for coding support with common languages (Python, C, JavaScript) and on common topics (web interfaces, databases, GUI design), it's extremely unlikely that you'll encounter hallucinations.

You can find some tips on the internet on how to recognize hallucinations. In our opinion, however, these recommendations aren't very effective. Test the code! If this doesn't work due to nonexistent functions, modules, and so on, search the web for the function names in the traditional way. This usually settles the case. You can ask the AI tool of your choice to make improvements, but if there is insufficient training material, the chances of functioning code aren't particularly high, even on the second or third attempt.

The following example was created using ChatGPT 3.

Prompt: In a Bash script, I use PHP to generate a bcrypted hash code of a password:

```
PW='topSecret'
echo "<?php echo password_hash('$PW',
                        PASSWORD_DEFAULT);" | php
```

Is there a way to generate the hash code without calling PHP?

Yes, you can generate the bcrypt hash of a password using the bcrypt command-line tool, which is available on most Unix-like systems.

The syntax to generate a bcrypt hash code of a password using the bcrypt tool is:

```
bcrypt <password>
```

[more details]

The problem is—there is no bcrypt command on Linux. Current ChatGPT versions are smarter and recommend the use of the widely used htpasswd command.

Yes, you can generate bcrypt hash codes without calling PHP by using other tools that support bcrypt hashing. One such tool is htpasswd, which is part of the Apache HTTP Server utilities. Here is how you can use it:

```
sudo apt-get install apache2-utils
htpasswd -nbBC 10 user "$PW" | cut -d ":" -f 2 [more details]
```

13.2.3 Logic Errors

One of the worst errors we encountered while working on this book was a mathematically incorrect optimization of code. The son of one of the three authors had found a Python script on GitHub that calculates the behavior of a mountain bike shock (*four bar linkage suspension*). This script uses the numpy and scipy modules to perform various calculations that neither father nor son really understood, but whose graphical visualization looked correct. Without any real understanding of the mathematical principles, it was possible to adapt the script to the specific requirements with the help of AI. So far, so good.

However, the resulting code was quite confusing. The AI tool will probably be able to improve that, right?

Prompt: Is there a way to further optimize the code?

[270 lines of Python code]

Yes, there are several ways to optimize the code further. We can focus on improving readability, reducing redundancy, and enhancing efficiency. Here's a refactored version of the code with these improvements:

- **Minimize Redundant Calculations:** Avoid recalculating the same values multiple times.
- **Use Functions for Repeated Code:** Create functions for repeated code blocks.
- **Vectorize Operations:** Where possible, use vectorized operations with NumPy to improve performance.
- **Enhance Readability:** Use more descriptive variable names and add comments for clarity. Here's the optimized code:

[200 lines of Python code]

The AI tool, in this case it was ChatGPT 4o, makes an honest effort to restructure the code, and it does a lot of things right. In particular, it was possible to swap out a code block that occurs several times to a function. This section of code was then shorter and easier to read. Perfect!

However, ChatGPT didn't stop at the mathematical functions of the program and also tried to simplify their code, for example:

```
# Jacobian of vector loop function, before optimization
def J_4bar(x, l1, l2, l3, l4, theta1, theta4):
    theta2 = x[0]
    theta3 = x[1]
    J = np.zeros((2, 2))
    J[0, 0] = l1 * np.cos(theta1) - l2 * np.sin(theta2) +
              l3 * np.cos(theta3) + l4 * np.cos(theta4)
    J[0, 1] = l1 * np.cos(theta1) + l2 * np.cos(theta2) -
              l3 * np.sin(theta3) + l4 * np.cos(theta4)
    J[1, 0] = l1 * np.sin(theta1) + l2 * np.cos(theta2) +
              l3 * np.sin(theta3) + l4 * np.sin(theta4)
    J[1, 1] = l1 * np.sin(theta1) + l2 * np.sin(theta2) +
              l3 * np.cos(theta3) + l4 * np.sin(theta4)
    return J

# Jacobian of vector loop function, after optimization
def J_4bar(x, l1, l2, l3, l4, theta1, theta4):
    theta2, theta3 = x
    return np.array([
        [-l2 * np.sin(theta2), -l3 * np.sin(theta3)],
        [ l2 * np.cos(theta2),  l3 * np.cos(theta3)]
    ])
```

The new code is undoubtedly much shorter, but some of the original parameters ($l1, l4$, theta1, and theta4) are no longer included in the calculation. Depending on the values of these parameters, the result of the optimized function may be completely wrong. It remains a mystery how ChatGPT came up with this "optimization."

What else went wrong with this example?

- The (mathematical) knowledge was insufficient not only for ChatGPT but also for its users. It's never a good idea to ensure the plausibility of the AI results if your experience or expertise is insufficient.

- The script to be optimized was already quite large at approximately 10 KB, at the limit of what common language models (as of the time of writing) can still process well in their entirety. It would have been safer to optimize parts of the code piece by piece.

- The instruction "optimize this code" was rather unspecific. *What* should be optimized? The performance? The functionality? The structure and readability of the code?

13.2.4 Security Issues

Perhaps the greatest danger comes from code that works but has inherent security issues. Such code isn't noticeable because it successfully passes the usual functional tests.

The underlying problem is that the internet is teeming with code snippets that have little regard for security. Such code flows into the training material of language models in the same way as "good" code in terms of security. On average, AI tools deliver equally good or equally bad code as can be found on the internet or in other training sources.

To be fair, it must be said that the use of AI tools does little to change whether a program is secure or insecure. Even many developers or programmers who are brilliant in their specialist areas lack an eye for safety implications. Ultimately, you need to have your code reviewed by security experts, whether written manually or with the help of AI. You must verify its behavior for incorrect or manipulated inputs using fuzzy tests and other methods.

We also want to illustrate the security issue with an example. As a starting point for a REST API that we wanted to create in Python, we asked ChatGPT to create a minimal Docker setup. We've shortened the explanations in the AI response.

> **Prompt:** I want to have a minimal setup to test a REST API running in a Docker container.
>
> 1. Apache or Nginx as proxy
> 2. Docker container (Python) with hello world API, GET / should return { "hello": "world"}
>
> As little code as possible. No setup for HTTPS/SSL keys (perhaps later).

To set up a minimal environment for testing a REST API running in a Docker container, we'll use Python with Flask for the API and Nginx as the proxy.

Step 1: Create the Flask API. Create a directory for your project, e.g., hello-world-api. Inside this directory, create a file named app.py with the following content:

```
from flask import Flask, jsonify


app = Flask(__name__)
@app.route('/')
def hello_world():
    return jsonify({"hello": "world"})


if __name__ == '__main__':
    app.run(host='0.0.0.0', port=5000)
```

Step 2: In the same directory, create a Dockerfile:

```
# Use the official Python image from the Docker Hub
FROM python:3.9-slim
# Set the working directory
WORKDIR /app
# Copy the current directory contents into the container
# at /app
COPY . /app
# Install Flask
RUN pip install Flask
# Make port 5000 available to the world outside this container
EXPOSE 5000
# Run app.py when the container launches
CMD ["python", "app.py"]
```

Step 3: Create a docker-compose.yml file in the same directory:

```
version: '3'
services:
  web:
    build: .
    ports:
      - "5000:5000"
  nginx:
    image: nginx:alpine
    ports:
      - "80:80"
    volumes:
      - ./nginx.conf:/etc/nginx/nginx.conf
    depends_on:
      - web
```

Step 4: Create a nginx.conf file in the same directory with the following content:

```
events { }
http {
  server {
    listen 80;
    location / {
      proxy_pass http://web:5000;
      proxy_set_header Host $host;
      proxy_set_header X-Real-IP $remote_addr;
      proxy_set_header X-Forwarded-For
        $proxy_add_x_forwarded_for;
      proxy_set_header X-Forwarded-Proto $scheme;
    }
  }
}
```

Step 5: Navigate to your project directory in the terminal and run the following command to build and start your containers:

```
docker-compose up --build
```

Step 6: Open your browser and navigate to *http://localhost*. You should see the JSON response:

```
{
  "hello": "world"
}
```

This minimal setup ensures that your REST API is running in a Docker container with Nginx acting as a reverse proxy. You can later extend this setup to include HTTPS/SSL if needed.

Basically, the answer is great for *fast prototyping*! Assuming a functioning Docker environment, the Hello World API runs in five minutes and provides a perfect starting point for further experiments. So, what's the problem? Unfortunately, there isn't just one, but rather an entire host of security issues:

- ChatGPT has opted for the totally outdated Python 3.9 (step 2). Version 3.12 was the current version at the time, and the next version 3.13 was about to be completed. For security reasons, you should always use the latest possible software versions.

- Three additional lines in Dockerfile are sufficient for the Python program to run with the REST API in the Docker container without root rights (also step 2). The improved Dockerfile would look like this (with comments on the changes):

```
FROM python:3.12-slim
# Create custom bentzer and your own group
RUN groupadd -r appgroup && useradd -r -g appgroup appuser
WORKDIR /app
```

```
COPY . /app
# Change access rights for app directory
RUN chown -R appuser:appgroup /app
RUN pip install Flask
EXPOSE 5000
# Run Python in the appuser account (instead of root)
USER appuser
CMD ["python", "app.py"]
```

- Instead of `docker-compose.yml` (step 3), the file name `compose.yaml` is now commonly used. The real problem here, however, is port 5000, which is only used for internal communication between the two services. `ports: 5000:5000` opens the port to the whole world. This setting would already be much safer:

```
# in docker-compose.yaml use port 5000 internally only
ports:
  - "5000"
```

A setup with a separate internal network would be ideal. This would require a few additional lines in `compose.yaml`.

- Port 80 is also opened in `docker-compose.yaml` so that the REST API can be tested. As long as the REST API is only intended for testing purposes (which is clearly the case here), it would be safer to restrict the use of port 80 to `localhost`. This means that the test system can only be used on the local computer, but not from outside via the network.

```
# in docker-compose.yaml allow only local access to
# port 80
ports:
  - "localhost:80:80"
```

That was a lot of Docker-specific details, which this book isn't about. Rather, the example is intended to make it clear that a perfectly functioning solution provided by AI tools is by no means automatically secure. Not only does this statement apply to Docker setups, but to any type of code or IT instructions that ChatGPT and others spit out. As long as you try out an application locally, security considerations usually only play a subordinate role. However, proper security auditing is essential before productive use! In our opinion, it's even better to pay attention to safety right from the start.

Our Own Fault

We were partly to blame for the fact that the solution provided by ChatGPT was inadequate in terms of security. We've explicitly asked for *as little code as possible*. But we could also have added the sentence: *Please provide a secure setup that can be adapted later for productive usage*. Neither ChatGPT nor other AI chatbots will then provide the perfect setup, but the number of security flaws will drop noticeably.

13.2.5   Less Than Ideal Solutions

There are a number of scientific studies that assess the quality of coding responses using established reference solutions. We particularly liked "Is Stack Overflow Obsolete?" by Samia Kabir et al. (May 2024). It compares ChatGPT's answers to popular coding questions with the top-rated solutions on Stack Overflow. For this purpose, ChatGPT 3.5 was used. The PDF of the paper and the underlying data (including all questions) can be found here:

- *https://dl.acm.org/doi/pdf/10.1145/3613904.3642596*
- *https://github.com/SamiaKabir/ChatGPT-Answers-to-SO-questions/tree/main*

The results of the paper are contradictory:

- **High error rate**
  With more than 500 prompts, ChatGPT's responses contained errors in 52% of cases. This doesn't mean that the answers were completely wrong, but they weren't completely right either. We've chosen the term "less than ideal" in the title of this section.

  As many as 77% of the responses contained redundant or superfluous information.

- **ChatGPT doesn't understand the question**
  The reason for the incorrect answers was often that ChatGPT didn't understand the question correctly.

- **Fewer errors with popular topics**
  Answers are more likely to be correct if established knowledge comes into play. The probability of errors increases with questions about new technologies or versions and with unpopular, rarely asked questions.

- **ChatGPT's answers are more digestible**
  Despite the higher error rate, two thirds of all developers prefer the ChatGPT answers to the solutions on Stack Overflow! This is because the answers are often more coherently argued and formulated in a more polite and positive way.

Needless to say, we want to add our own unscientific stuff to this carefully researched publication:

- **Questionable error rate**
  The error rate seems very high to us. Of course, all language models make mistakes—that's what this chapter is all about. But we haven't experienced that every other answer would be wrong. Perhaps this is because we've been working with the next version of GPT throughout. (By the time you read this book, the generation after the next one may already be here!)

- **Functioning, but not optimal code**
  In our work, the code from ChatGPT often worked, so it wasn't wrong. But it was often possible to improve the code, for example, by using modern/newer functions

or by applying more elegant algorithms. The AI code wasn't wrong, but it wasn't always as perfect as good quality code written by human developers. (The benchmark would be code from an experienced developer or a well-trained programmer with several years of training and experience.)

- **Targeted response**
  We see the biggest advantage of ChatGPT or other language models compared to Stack Overflow in the fact that AI tools respond very specifically to our individual question. In practice, it's relatively rare for a Stack Overflow article to answer your exact question. In fact, a Stack Overflow search often comes up with three or four articles that answer partial aspects of the question. Based on that, we then more or less laboriously put together our own code.

  The authors of the study turned the tables for the investigation and incorporated established Stack Overflow questions directly into the prompt. That seems a little unfair to us.

  In other words, regardless of all the mistakes the AI tool makes, it still answers exactly the question asked (assuming the wording of the prompt is sufficiently clear). An internet search, whether on Stack Overflow or elsewhere, will find countless more or less relevant information in the wider context of the question.

- **Less would be more**
  We share the criticism of the paper authors that ChatGPT's answers are almost always too lengthy. We don't want to read the solution three times—in the introduction, in the code, and then again in the explanations that follow!

  Admittedly, whether the length of an answer is appropriate or not depends very much on your own level of knowledge. The less prior knowledge, the more helpful the redundancy of the AI text. Personally, we like the AI tool *Claude* from Anthropic better than ChatGPT in this respect: Claude gets to the heart of the matter and is far less communicative.

- **Commercial language models are the benchmark**
  In contrast to the paper authors, we have not only used ChatGPT in this book but also tried other language models (often smaller LLMs executed locally with Ollama). Although we would prefer it if we could write the opposite here: the commercial cloud offerings are more convincing in terms of quality.

Further Reading

There are various sites on the internet that compare the quality and error-proneness of language models during code generation:

- *https://symflower.com/en/company/blog*
- *https://artificialanalysis.ai/leaderboards/models*
- *https://github.com/continuedev/what-llm-to-use*

In our view, the results of the annual Stack Overflow survey (*https://survey.stackoverflow.co/2024/ai#efficacy-and-ethics*) on AI tools are also very interesting. Although three quarters of all developers already use AI tools or want to do so in the near future, the developer community sees two fundamental problems: 66% of survey participants generally have doubts about the quality or accuracy of AI answers, and 63% see the main problem with AI tools as being that they lack an overview of the entire code base (more precisely: *AI tools lack the context of the code base, internal frameworks, and/or company knowledge*).

13.3 Ethical Issues

Can the use of AI systems for software development be approved of from an ethical point of view? This question is perhaps asked a little late because we've almost reached the end of this book. The fact is that AI tools have more downsides than the occasional wrong answer. That is why we want to deal with this in the following sections.

Restricting the Use of AI for Coding

We've already mentioned this in the preface and would like to point it out again here: this book deals with AI exclusively in the context of software development.

Of course, we know that AI tools are used to generate fake news, images, and videos, and that AI algorithms have the potential to put entire professional groups out of work. AI is finding its way into more and more areas of our lives. The resulting problems will fundamentally change our society. These problems are undisputed, but like any AI application outside of coding, they aren't the subject of this book.

The same applies to the dystopian question of whether AI models will soon evolve into a superintelligence and then take over the world. This is still the stuff of science fiction books and films. The only worrying thing is how many renowned AI researchers see this as a real danger (read *https://righttowarn.ai*).

13.3.1 The Origin of AI Knowledge

Each of us three authors had already written other books before this one. Time and again, we've asked ourselves the following question: Were our books also used to train the language models whose application we describe here? Have our posts in discussion forums been used, our code stored on GitHub, the articles in our blogs? These questions are difficult to answer. Most AI providers are very tight-lipped when it comes to a precise description of the training material.

Another question, however, is very easy to answer: What did we get in return for using our knowledge—in whatever form—to train LLMs? Definitely nothing!

Microsoft, which in the past has always demanded very strict compliance with all conceivable (and inconceivable) rules on copyright and licensing issues, has surprisingly liberal ideas when it comes to AI training. Mustafa Suleyman, CEO for AI issues at Microsoft, put it this way: "I think that with respect to content that's already on the open web, the social contract of that content since the '90s has been that it's fair use. Anyone can copy it, recreate with it, reproduce with it. That has been 'freeware,' if you like, that's been the understanding" (6/26/2024). Naturally, this statement was met with a lot of criticism. Ultimately, however, Mr. Suleyman has only expressed what other AI companies are probably thinking to themselves.

There is no question that fair compensation for training material consisting of billions of pages of text and terabytes of data is virtually impossible. Who should get how much? How can the contribution of a text to the overall language model be understood? Tracing every single author would involve a huge amount of work. Think of sites like Stack Overflow with contributions from millions of developers or articles on Wikipedia that many authors have contributed to. But if you follow Microsoft's way of thinking, Stack Overflow is a great freeware knowledge base for AI training. At the same time, the business model of Stack Overflow is being shaken by AI offerings. Is that "fair use"?

The difficulties in realizing fair knowledge compensation don't change the fact that the entire AI business model has a massive downside. On one hand, there are IT companies worth billions that earn money with AI technology. On the other hand, there is the whole of humanity, which has created this knowledge.

Admittedly, many AI offerings on the web are available free of charge, albeit with certain restrictions. Nevertheless, we find it irritating that our code stored on GitHub is used for AI training, while at the same time we have to pay again to use the model (also) trained with our code in GitHub Copilot.

There is no doubt that AI research, the training of AI models, and so on cost a lot of money. AI companies need a way to earn money, or further development is difficult. So far, in our capitalist world view, everything is somehow in order. It remains to be seen whether part of the revenue shouldn't flow back to where the knowledge was created or made accessible (schools, universities, publishers, etc.).

Sooner or later, courts will have to decide which data may be used for AI training and under what conditions. A legal dispute has already broken out over the use of GitHub code for AI training. As of December 2024, it looked as if the judges would side with Microsoft (read *www.developer-tech.com/news/judge-dismisses-majority-github-copilot-copyright-claims*).

13.3.2   Jobless IT Experts

Microsoft advertises GitHub Copilot with "55% faster coding." If this were true, a company could lay off 35% of its software developers and give the rest a Copilot subscription. The downsized team would produce the same amount of code as before.

This calculation is misleading for various reasons, not least because software developers aren't just responsible for coding. They take on many other tasks where AI support is less or not at all relevant. Last but not least, this includes (human) contact with team members and customers.

NVIDIA's boss Jensen Huang is unimpressed. He goes even further and believes that software developers will be completely superfluous in the future. Of course, this statement is primarily due to marketing. Nevertheless, all IT experts must ask themselves whether they are making themselves partially redundant by using AI technologies. As authors, we should ask ourselves: Are we supporting this process with this book?

The Dead Live Longer

The risk of IT hordes seeking new jobs is definitely not dramatic, at least not as a result of the new AI tools. They aren't yet good enough for that. And even for the near future, we're skeptical that AI tools will replace a lot of real IT jobs. The following blog article sums up this view convincingly: *https://dava.ai/n/hank/ai-didnt-take-my-job*.

However, we expect AI tools to massively change work in many IT areas. Well-trained, experienced developers will definitely have an advantage on the job market; ideally, they should be able to use AI tools efficiently.

What is interesting in this context is that managers have different, often completely exaggerated expectations than developers who actually work with AI tools. According to a study by Atlassian (*www.atlassian.com/software/compass/resources/state-of-developer-2024*), only just under 40% of respondents believe that current AI tools significantly improve their productivity. However, 60% still expect AI tools to significantly improve their productivity within the next two years.

In 2024, the annual Stack Overflow survey (*https://survey.stackoverflow.co/2024/ai#efficacy-and-ethics*) mentioned earlier in this chapter found out for the first time whether IT experts fear for their jobs because of AI tools: 68% don't see this risk, while 12% do (the rest are unsure).

13.3.3   Unlimited Energy Consumption

Cloud-based AI services start answering in a fraction of a second and present their answers faster than you can read along. You don't realize how much energy this requires until you try to run language models locally. The MacBook of one of the authors of this book was practically permanently idle during work. This only changed when Ollama had to execute requests to a language model locally: the power consumption immediately increased from 3 watts to 20 to 30 watts (in each case without screen, the MacBook was connected to an external monitor). However, these are still modest values compared to the power consumption of a "classic" PC with an NVIDIA graphics card: Although the computing power is then also sufficient for significantly faster processing of larger voice models, the power consumption increases to several hundred watts. That's crazy!

The big AI companies will only smile wearily at the figures just presented. The data centers for running existing AI offerings and training new voice and multimedia models require as much power as entire small towns. Thanks to new hardware and improved optimization, it's hoped that the power consumption per prompt or for training new models will fall in the coming years. This is offset by increasing quality requirements and user numbers. Overall, it's to be feared that the demand for electricity for AI applications will continue to rise in the coming years.

Of course, every other IT service also needs electricity: the local operation of notebooks, monitors and Wi-Fi routers; storing data in the cloud; streaming and playing a video; and so on. But the power consumption of AI services goes beyond all previously known proportions. In view of the climate crisis, it's fair to ask to what extent AI services are sustainable or can become so in the future.

13.4   Conclusions and Outlook

Now that we've compared the pros and cons in detail, we present our conclusions and our expectations for the coming years in the final section of this book.

13.4.1   Better Results with More Prior Knowledge and Experience

Reporting on AI tools sounds something like this: "Two or three prompts, and the new web application is all set!" This completely overlooks the fact that the efficient use of AI tools only works with prior knowledge. If you've already created a dozen web applications, have fallen into a couple of traps, and are familiar with the usual libraries and tools, then you can easily follow the statements of ChatGPT and others. You can separate the wheat from the chaff in AI-generated code at a glance. However, if you program your first web application, AI tools can at best point you in the right direction. The road remains rocky. The end product won't match the quality of a program that experienced developers would create.

Because the answers from AI tools are always elegantly and politely formulated, it's difficult to distinguish between correct and incorrect answers. It's also difficult to predict in advance which questions the AI tool will excel at and where it will fail.

In other words, AI tools are no substitute for IT training or real-life experience. The more you know or can do yourself, the more AI tools will help you.

The More Incompetent, the More Self-Confident

In psychological literature, the Dunning-Kruger effect (*https://en.wikipedia.org/wiki/Dunning%E2%80%93Kruger_effect*) describes the overestimation of one's own competence by incompetent people; that is, newcomers often greatly overestimate their own knowledge.

This effect applies exactly to AI! If you assign tasks to AI where you're not competent yourself, you can't assess whether the solutions are reasonable.

13.4.2   Cloud or Local AI?

When we looked more closely at the local execution of language models, we were thrilled. In our view, there are many arguments in favor of executing AI functions for coding locally: data protection, independence from commercial AI providers, costs, and so on.

The "local" execution of the language model doesn't necessarily have to take place on your own computer. A good alternative is a dedicated server set up within the company or in a data center for AI tasks.

However, there are several reasons that speak against the "local is better" or at least "local is safer" approach:

- **Quality**
  In our tests, commercial language models such as GPT-4o were clearly superior to the free models. This is not only because these models as such are better. For local execution, you're often forced to use small variants of a free language model due to limited storage space. A local 7b or 15b model can't compete with a model with several hundred billion parameters.

- **Money**
  AI companies understandably want to earn money. The easiest way to do this is with cloud subscriptions. For this reason, companies such as OpenAI have no interest in you switching to open-source software and models. They will do anything to ensure that their own commercial AI offering is better than the free competition. And because AI research and training costs a lot of money, it's conceivable that the gap between commercial and free AI tools will widen rather than narrow.

- **Lack of local hardware**
  Most computers and notebooks currently on sale (as of early 2025) have far too little power to run language models locally with sufficient quality and speed.

  That will probably change. We expect AI-compatible hardware to become more established. This hardware is characterized by more RAM, higher GPU performance,

13

and the integration of dedicated *neural processing units* (NPUs). AI research is also intensively searching for ways to execute language models more efficiently (new algorithms, more space-saving models, etc.).

Nevertheless, your own laptop or a local AI server will still not be able to keep up with the hardware infrastructure of AI cloud providers in the future.

Based on these points, it can't be ruled out that the cloud approach will prevail—at least for the next few years. An astonishing number of companies already see no problem in storing all of their data in the cloud (Microsoft 365 and others). It's quite possible that this trust will also extend to AI cloud services in the future.

Risk of a Two-Tier IT Society

The majority of our team of authors is at home in the open world. We appreciate that the Linux distribution of our choice, programming languages, developer tools (Docker, Git, etc.), editors, and IDEs are freely available.

Are AI tools throwing a spanner in the works here? As long as free language models aren't competitive or as long as their local execution requires hardware resources that aren't available to private users, there is almost no way around paid AI subscriptions.

You then have the choice: local execution of free language models versus commercial AI cloud offerings. If you can afford it, or if you have no problem with cloud tools seeing all your code, you'll get better AI support than with free, local tools without privacy concerns. We think this is a worrying development.

13.4.3   Quality Enhancement through Training with AI-Generated Code

The first generation of language models was trained exclusively with texts and code written by humans. The more developers use AI tools, the more code repositories, blogs, and online manuals contain code written partly or entirely by AI tools.

Future language models are therefore trained with code produced by a preceding version. To put it even more pointedly: AI trains itself. It's not yet clear how this will affect future language models.

13.4.4   The Coming Few Years

Despite the training problem just mentioned, we expect that future language models will work better than the versions presented in this book. New neural algorithms, new training methods, and new hardware will help.

The question is rather whether AI support for coding will improve in small incremental steps or whether there will be further breakthroughs as in previous years. Of course, we can't say for sure, but we're expecting a somewhat calmer development. From the perspective of this book, our wish list looks something like this:

- **LLMs optimized for coding**
  Current language models are *also* able to program. This ability is more of a side effect that results from the way language models work. What we want to see, however, is language models that are designed specifically for coding and that are better suited to the automation of coding tasks. At this point, we've repeatedly come up against the limits of current language models in this book (see in particular Chapter 10 and Chapter 11).

- **More context**
  From an IT perspective, the largest possible context window is desirable so that the language model can ideally take into account all the existing code in a project.

  In fact, the size of the context window has recently increased in many LLMs, but at the expense of performance. The more context, the slower the response generation. In our experience, the quality of AI responses also decreases in current models with too much context. To put it casually, too much, too broad a context confuses the model; the answers are less targeted and sometimes have little to do with the question. There is still a lot of room for improvement here.

- **Local LLMs**
  Despite all the current limitations, we see a great future in the local execution of language models. Perhaps it will be possible to reduce the size of language models (i.e., to focus on the content that is essential for coding). This would make it possible to achieve a higher execution speed without any loss of quality. The effort to make LLMs smartphone-compatible could help here. If in the future, a smartphone is sufficient to run a local language model satisfactorily, then the same should all the more be the case for a powerful developer notebook.

We're aware that current AI tools weren't primarily created for coding. The focus of further development is also on where the most money can be made, and that is certainly not the comparatively small software developer community. In this respect, we should be a little skeptical as to whether our wishes will come true.

13.4.5   Conclusion

This brings us to the end of this book. Finally, here are four theses:

- **AI tools have become an integral part of coding**
  For many developers, the use of AI tools for coding is already just as much a matter of course as the use of the spell-check function in Word or other word processors. This isn't perfect, but useful.

- **Use only within the scope of your own knowledge**
  AI tools increase coding efficiency, but can't replace your own basic knowledge. We consider the use of AI-generated code that you don't understand to be highly negligent.

- **Errors keep occurring**
  We expect the error rate to decrease somewhat over time, but we're sure that AI tools will continue to produce errors in the future. This is precisely why it's so important that you understand AI-generated code and take the trouble to test it thoroughly.

- **Seize opportunities and don't dread new developments**
  In Europe, we look at the risks of new technologies first and look for regulatory mechanisms. In the United States, it tends to be the other way around, with enthusiasm prevailing. It's certainly debatable which path makes more sense, but it would definitely be a shame to overlook the opportunities that are currently opening up in the AI sector.

The Authors

Michael Kofler is a programmer and one of the most successful and versatile computing authors in the German-speaking world. His current topics include AI, Linux, Docker, Git, hacking and security, Raspberry Pi, and the programming languages Swift, Java, Python, and Kotlin. Michael also teaches at the Joanneum University of Applied Sciences in Kapfenberg, Austria.

Bernd Öggl is an experienced system administrator and programmer. He enjoys experimenting with new technologies and works with AI in software development using GitHub Copilot. He is particularly interested in local large language models (LLMs) and advancements such as OpenHands and retrieval-augmented generation (RAG).

Sebastian Springer is a JavaScript engineer at MaibornWolff. In addition to developing and designing both client-side and server-side JavaScript applications, he focuses on imparting knowledge. He inspires enthusiasm for professional development with JavaScript as a lecturer for JavaScript, a speaker at numerous conferences, and an author. Sebastian was previously a team leader at Mayflower GmbH, a premier web development agency in Germany. He was responsible for project and team management, architecture, and customer care for companies such as Nintendo Europe and Siemens.

Index